About Island Press

Since 1984, the nonprofit Island Press has been stimulating, shaping, and communicating the ideas that are essential for solving environmental problems worldwide. With more than 800 titles in print and some 40 new releases each year, we are the nation's leading publisher on environmental issues. We identify innovative thinkers and emerging trends in the environmental field. We work with world-renowned experts and authors to develop cross-disciplinary solutions to environmental challenges.

Island Press designs and implements coordinated book publication campaigns in order to communicate our critical messages in print, in person, and online using the latest technologies, programs, and the media. Our goal: to reach targeted audiences—scientists, policymakers, environmental advocates, the media, and concerned citizens—who can and will take action to protect the plants and animals that enrich our world, the ecosystems we need to survive, the water we drink, and the air we breathe.

Island Press gratefully acknowledges the support of its work by the Agua Fund, Inc., The Margaret A. Cargill Foundation, Betsy and Jesse Fink Foundation, The William and Flora Hewlett Foundation, The Kresge Foundation, The Forrest and Frances Lattner Foundation, The Andrew W. Mellon Foundation, The Curtis and Edith Munson Foundation, The Overbrook Foundation, The David and Lucile Packard Foundation, The Summit Foundation, Trust for Architectural Easements, The Winslow Foundation, and other generous donors.

The opinions expressed in this book are those of the author(s) and do not necessarily reflect the views of our donors.

ENERGY DEVELOPMENT AND WILDLIFE CONSERVATION IN WESTERN NORTH AMERICA

Energy Development and Wildlife Conservation in Western North America

Edited by
David E. Naugle

Foreword by Mark S. Boyce

Washington | Covelo | London

Library of Congress Cataloging-in-Publication Data

Energy development and wildlife conservation in western North America / edited by David
E. Naugle ; foreword by Mark S. Boyce.
p. cm.
Includes bibliographical references and index.
ISBN-13: 978-1-59726-657-4 (hardcover : alk. paper)
ISBN-10: 1-59726-657-4 (hardcover : alk. paper)
ISBN-13: 978-1-59726-658-1 (pbk. : alk. paper)
ISBN-10: 1-59726-658-2 (pbk. : alk. paper) 1. Energy development–Environmental
aspects—West (U.S.) 2. Energy development—Environmental aspects—Canada, Western.
3. Wildlife conservation—West (U.S.) 4. Wildlife conservation—Canada, Western.
I. Naugle, David E.
TD195.E49E528 2011
333.79′150978—dc22
2010029433

Text design and typesetting by Karen Wenk

Printed on recycled, acid-free paper

Manufactured in the United States of America

10 9 8 7 6 5 4 3 2 1

Key Words
Conservation planning, cumulative effects, energy development, human footprint, oil and
gas, renewable energy, sage-grouse, wildlife conservation, wind power, woodland caribou,
climate change, energy demand, pronghorn, endangered species, renewable energy, biofuels,
solar energy, conservation in the American West, community-based landscape conservation,
oil, natural gas, hydrocarbons, tar sands.

CONTENTS

MARK S. BOYCE

Most of western North America is ranchland. Ranches are large, and some ranchers are powerful people in state and provincial politics. Until recently, the best and most profitable use of the land was thought to be for raising cattle. But the livestock industry has shriveled such that livestock production accounts for only 1 percent of the gross domestic product in western states and provinces. In contrast, energy development has soared with oil and gas drilling; surface mining of oil sands, oil shale, and coal; and underground gasification, steam-assisted gravity drainage, and coal bed methane production. Revenues and royalties are so great from energy extraction that environmental regulations are often compromised to ensure that energy resources are developed.

Along with the changing economies of the West have come new challenges to managing wildlife in areas affected by industrial development. Most egregious to me is the erosion of habitats for threatened and endangered species that continues despite legislation that is supposed to protect these species. The demand for energy resources is so great and the economic values are so large that environmental considerations often are ignored.

These habitat losses are particularly disturbing when one realizes that minor changes by industry could substantially reduce the footprint associated with energy development. Directional drilling, seasonal surface occupancy, gating of roads, and minimized vegetation disturbance can be highly effective at reducing the effects of energy development on wildlife. Canadian Forces Base Suffield demonstrates how gas development can be nearly invisible without permanent roads and underground wellheads and pipelines. In some cases corporate social responsibility results in best management practices by the energy developers, but in remote areas with low human populations corporate responsibility to shareholders often prevails. Responsible energy development can be guaranteed only with sound environmental regulations and government oversight.

Government oversight is necessary to coordinate multiple users of the land. The cumulative effects of energy development, along with timber harvest, agriculture, and recreation, can be destructive. Coordinating development by many stakeholders can be challenging. The province of Alberta is developing a land use framework to tackle this problem, and various land management agencies throughout the West have permitting requirements intended to monitor and coordinate development.

Science has much to offer to industry and government to assist with sustainable development. Advances in geographic information systems make it fairly easy to anticipate the consequences of alternative development plans. We can map the distributions of species, and we can project how these distributions will change given alternative future landscapes. By running alternative scenarios in a computer, we can find the best solution for development that will have the least impact on the environment.

Although such scenario modeling is possible, without background research and data such scenarios are fantasy. Unfortunately, we often do not know the consequences of development. Likewise, we do not know how various management practices will affect wildlife. There is opportunity to join forces with industry and government to design experiments so that researchers can document the consequences of alternative development schemes. As we accumulate results from such studies, we can improve our ability to make reliable predictions of the consequences of land use practices. Such joining of forces would be a fantastic step forward, and I have worked on three such efforts, each of which has failed.

Ultimately, scientists have visions of development being shaped by adaptive management. This entails predicting the consequences of development based on scenario modeling, implementing a development or a change in management, monitoring the consequences, and revising the model so that we might do a better job of anticipating the consequences of the next development. The idea is to inject the scientific method into the process of natural resource management. In practice, however, adaptive management is rare. Seldom is scenario modeling done before development, and seldom is monitoring in place with sufficient precision and replication to document the consequences. Implementation requires that someone perform the modeling and collect the monitoring data, which costs money. If modeling and monitoring are not required by government regulations, it is unlikely that anyone will take the initiative. Some resource managers are not supportive of adaptive management because research and monitoring might yield data that suggest that a change is needed; it is easier to maintain the status quo.

Even though the energy sector makes vast amounts of money from the extraction of fossil energy resources, there are few incentives for investment in environmental programs and wildlife conservation. Investment in high-profile wildlife projects might be supported if it can improve corporate public image, but conservation investments are most likely to happen if encouraged by regulations or permitting requirements. Conservation offsets are sometimes leveraged to purchase land for conservation as a trade for land destroyed by development (e.g., oil sands and strip mining).

Energy use from fossil fuels is rapidly changing global climates, with potentially disastrous consequences for the future of agriculture and conservation. An offshoot of recent concerns about carbon emissions is the potential to manage conservation properties for carbon sequestration. Grasslands can be highly effective places to sequester carbon in western North America because the carbon is safely underground, where it might stay if the land remains in continuous grass cover. In contrast, forests in the North and in the Rocky Mountains burn at various intervals, discharging the carbon back into the atmosphere. Taxes on corporate carbon emissions could be invested in conservation projects allowing plants to sequester carbon into the soil. Many energy extraction industries have vast carbon emissions; for example, steam-assisted gravity drainage and oil sand operations burn natural gas to separate the oil from the sand. Engineering solutions for carbon capture and storage can be very expensive relative to using natural grassland vegetation to sequester carbon from the atmosphere.

Croplands in North America have lost 40–60 percent of the carbon from the soil, but when cropland is restored to native grass cover we observe a rapid rebound, with 20–30 metric tons of carbon dioxide equivalents being sequestered per hectare in the first 15–20 years. At a global scale, land use changes, such as converting marginal cropland to permanent grassland, easily can compensate for current carbon emissions.

I am optimistic that future landscapes of western North America will continue to support productive vegetation and thriving populations of wildlife. But to ensure that this occurs we must coordinate planning to reduce cumulative effects, apply best management practices, and pay special attention to key habitats. Examples reviewed in this book can show us the way forward.

PREFACE

To many people, the word *West* conjures up mental images of wide open spaces that support world-class populations of mule deer and iconic species such as pronghorn. But with increasing energy demands, open space is at a premium, and poorly placed developments threaten our wildlife heritage. From boreal caribou in Alberta to sage-grouse in Wyoming, much has already been lost, and recent increases in domestic production to reduce dependence on foreign oil portend the challenge ahead. And as this crisis deepens, decision makers look to science to help them develop solutions to maintain viable and connected wildlife populations before this conservation opportunity is lost.

Human demand for energy, which is projected to increase by 50 percent by 2030, is an issue of economic and national security in the United States and Canada. The question of increasing energy development in the West is not whether to do so, but rather where to reduce impacts and still extract resources to meet domestic demand. Halting development would result in economic hardship, yet too much development in places that support imperiled species will invoke federal laws that protect at-risk ecosystems. The key is to locate energy developments to reduce impacts on wildlife populations and other natural resources. Lawsuits and political wrangling will continue until we implement tangible on-the-ground conservation at scales equivalent to those of development.

This book provides a vision for landscape conservation that elected officials, industry representatives, natural resource managers, conservation groups, and the public can use to safeguard our wildlife heritage while securing our energy future. I conceived of this book in 2005 while conducting research in Wyoming's Powder River Basin, where we first documented the cumulative impacts of energy development on sage-grouse. Since then, although myriad studies have demonstrated cumulative impacts of energy development on populations of imperiled species in prairie, shrubland, and forested landscapes throughout the West, no book-length synthesis has been published. This void in the conservation literature at first seemed ironic because energy independence is a major issue that will be debated for years to come. Perhaps the void is best explained by its recent emergence; indeed, the peer-reviewed science on energy and wildlife impacts has been published in just the last 10 years. This book synthesizes the pertinent scientific information, and the Literature Cited is listed at the end of the book to reduce redundancy between chapters.

Tradeoffs between energy development and conservation are unfolding before our eyes, and the intention of this book is to help policymakers turn science into solutions to this pressing issue. The right science is a rallying point that allows conservation partners to focus on their similarities rather than their differences. This book speaks to a philosophy of science-based conservation that seeks to understand how a system works and then to use that knowledge to develop solutions. In part I, we frame the issue, describe the major types of extraction, and quantify the pace and extent of current and future development. In part II, we provide the biological foundation for understanding cumulative impacts, synthesize the biological response of wildlife to development, discuss energy infrastructure as a conduit for the spread of invasive species, and compare impacts of alternative energy with those of conventional development.

Finally, in part III we call for a paradigm shift away from random opportunism to broad-scale and strategic planning and implementation of conservation in priority landscapes. We show how science can help identify landscapes that support viable caribou populations, delineate core areas for sage-grouse, and forecast future development scenarios to aid in conservation design. We champion community-based landscape conservation as a solution for maintaining large and intact habitats that support healthy wildlife populations. Most importantly, we weave solutions into the social fabric of communities and rural ways of life, for the will of the people of the West, not its governments, will ultimately determine our conservation future.

We also provide readers with a photo essay of the major types of extraction and their associated footprints. Many readers have not experienced firsthand the density of roads, transmission and seismic lines, traffic, noise, and other forms of human disturbance that accompany development in otherwise small and traditional ranching communities. Readers unfamiliar with energy development may be surprised by the extent of the impact associated with extractive activities. Our photo essay illustrates the magnitude of cumulative impacts of energy development facing mule deer, pronghorn, caribou, sage-grouse, and other western icons. Aerial photos of development impress on the reader that impacts from an individual oil well or wind turbine pale in comparison to multiplicative impacts of development that accumulate across the broader landscape. The text that accompanies the photo essay conveys why the scale of conservation must be analogous to that of development if we are to maintain the large and open spaces on which wildlife depends.

ACKNOWLEDGMENTS

I am, first and foremost, indebted to the chapter authors who generously provided their particular expertise and insight. I am honored to work with such an enthusiastic group of talented people. Barbara Dean and Erin Johnson of Island Press are consummate professionals, and I appreciated their help every step of the way. I. Joseph (Joe) Ball, former leader of the Montana Cooperative Wildlife Research Unit, provided stylistic editing that greatly improved the manuscript.

My work on energy and wildlife issues is a journey along which I have met people who have inspired and influenced my thoughts about science and conservation. Joe Kiesecker and Holly Copeland of The Nature Conservancy continually challenged me to work at appropriately large spatial and temporal scales. Tom Rinkes of the Bureau of Land Management in Idaho and Dale Tribby of the Bureau of Land Management in Montana helped shape my thoughts on how multiple-use agencies balance energy and wildlife resources. Tom Christiansen with Wyoming Game and Fish and Rick Northrup and Jeff Herbert from Montana Fish, Wildlife and Parks inspired me to always make science applicable to management. Pat Fargey of Parks Canada and Joel Nicholson of Alberta Fish and Wildlife provided the Canadian perspective on energy development and conservation of natural resources. Greg Neudecker of the Partners for Fish and Wildlife Program (U.S. Fish and Wildlife Service) taught me the art of working with people; he will forget more about community-based landscape conservation than I will ever know.

My three former graduate students who worked on energy and wildlife issues have forever infected me with their passion for science. I thank Brett Walker for his skill in research design and for his unending quest to understand underlying mechanisms, Kevin Doherty for his quantitative aptitude and his knack for turning science into solutions, and Jason Tack for his uncanny ability to ask great questions. I was supposed to be mentoring you, but you were really teaching me; I wish you the best in your careers.

I thank numerous organizations for funding my energy and sage-grouse research, which immersed me in the topic and ultimately resulted in this book. I thank the Bureau of Land Management, the primary source of funding. Additional support came from National Fish and Wildlife Foundation; Wolf Creek Charitable Foundation (Bob Berry); Liz Claiborne Art Ortenberg Foundation; Petroleum Association of Wyoming; Western Gas Resources Incorporated; Bighorn Environmental Consulting; Grasslands National Park of Canada; World Wildlife Fund; Anheuser-Busch Companies Incorporated; Montana Fish, Wildlife and Parks; Wyoming Game and Fish Department; Montana Cooperative Wildlife Research Unit; and the University of Montana. I thank Perry Brown, dean of the College of Forestry and Conservation at the University of Montana, and Daniel Pletscher, director of the Wildlife Biology Program, for giving me the time and support to accomplish this task.

Lastly, I thank my wife, Corey, for her patient support of my science and conservation endeavors.

PART I

Energy Development and the Human Footprint

Forecasts point to the Rocky Mountain West as a primary place where the United States and Canada must look to increase energy production. The essence of the conflict between energy development and wildlife conservation in the West is the large amount of spatial overlap between competing resource values. Many of the landscapes being developed, and others that have been leased for exploration and potential development, overlie our largest and best remaining wildlife habitats. Viable solutions to this conflict must include large open spaces for wildlife because modern-day energy developments are industrial zones with disturbance levels that are incompatible with wildlife conservation. Wind and solar developments address climate change problems by reducing carbon emissions and reduce air and water pollution by providing clean and renewable energy. But simply switching a portion of our energy portfolio to renewable sources will not solve wildlife problems because wind and solar production requires an amount of space per unit of power second only to that needed for biofuel production. Rather, placing wind and solar developments in areas that have already been heavily disturbed by people will help us realize all the benefits of renewable sources of energy.

The first step toward sustainable development is an unbiased inventory and analysis of our onshore energy resources and a working knowledge of

1

the human footprint that accompanies development. Part I characterizes the increasing demand for energy and quantifies the extent to which major biomes will be affected by development.

Chapters 1 and 2 provide the big energy picture that is the background and foundation for the rest of the book. New and existing energy development may directly or indirectly affect 96 million hectares (291 million acres) of the five major biomes in western North America. Boreal forest, shrublands, and grasslands are especially vulnerable because of their geographic concurrence with the sedimentary basins that hold hydrocarbon deposits. These same systems will be further affected if renewable energy development proceeds on a maximum-development basis. Predicted impacts resulting from renewable energy extraction are especially disconcerting because the affected systems support high biodiversity yet have received little protection.

Chapter 1

Introduction to Energy Development in the West

DAVID E. NAUGLE AND HOLLY E. COPELAND

The story of North American "progress" is best characterized by the wave of human influence that originated in the East and spread westward. We first cleared eastern forests for European settlement and subsequently plowed midcontinent grasslands to produce food and fiber. Now the heavy footprint of energy development threatens to destroy the last of our large and intact western landscapes. People in the West are beginning to realize the social and economic tradeoffs associated with burgeoning development. Canadians enjoy the economic gains from exporting energy to U.S. markets but worry that declines in air and water quality accompanying extraction may be too high. Americans happily consume Canadian imports because buying oil from countries unfriendly to the United States poses a threat to national security. Energy development is a key to domestic prosperity in both countries, but poorly planned and largely unregulated, it comes at a high cost to nature.

We define the West as the eleven U.S. states located west of and including Montana, Wyoming, Colorado, and New Mexico and the Canadian provinces of Saskatchewan, Alberta, and British Columbia. Extracting oil, gas, coal, and uranium in the West is not new, but the pace and extent of development are. Also new is the realization that the West harbors some of the best renewable energy resources—plenty of wind, sun, and geothermal power—at a time when clean, green energy is part of a critical long-term solution to the problems of energy security, carbon emissions, and pollution.

The famous NASA nighttime Earth satellite image tells the story best. While we have settled the coasts and heartlands of North America, the interior West has remained largely dark. With the addition of new wind turbines, wells, and mines, we risk losing our last dark spaces on the map.

Since the late 1990s, as energy development intensified throughout the West, scientists began carefully studying this development and its effect on wildlife populations and ecosystems. In Canada, energy-related roads and seismic lines cut through the boreal forest have decreased populations of woodland caribou (*Rangifer tarandus tarandus*) through increased predation by wolves (*Canis lupus*) (chap. 5). In Montana and Wyoming, sage-grouse (*Centrocercus urophasianus*) populations are declining because adult birds remain in traditional nesting areas regardless of increasing levels of development, only to experience high rates of mortality, and yearlings that have not yet imprinted leave the gas fields in an attempt to escape human disturbance (chap. 4). In Wyoming, studies have shown that energy development has severed historic pronghorn (*Antilocapra americana*) migration corridors linking breeding and winter ranges (chap. 5). In addition to wildlife impacts, scientists are concerned that energy development acts as a conduit for invasive plant species, altering and degrading otherwise intact and functioning landscapes (chap. 7). Together, these studies implicate the cumulative effect energy development has on wildlife populations, resulting in declines of many iconic western species and the habitats on which they depend. These species are biologically important to the ecosystems that they inhabit and also socially relevant to the people who live and recreate in the West, resulting in heightened public awareness of impacts and an intensified desire to find a balanced solution to development.

Using less energy is an obvious and partial solution to the problem. Conservation efforts in the United States could reduce overall global demand because the United States consumes 21 percent of all the energy the world produces. To date, the systemic changes needed for significant energy conservation have not yet occurred, and projections in future U.S. energy demands by leading experts reflect this failure. Energy demands in the United States are projected to grow 0.5–1.3 percent annually (Energy Information Administration [EIA] 2009a). Projections that incorporate conservation-related energy policy changes, best available technology, and increased prices still indicate overall annual U.S. energy demand growing from 107.5 exajoules (1 exajoule = 0.95 quadrillion British thermal units) in 2007 to 115.8 exajoules in 2030, an increase roughly equivalent to California's current annual energy consumption. These projections show that conservation and energy efficiency measures could reduce overall residential demand by 1 percent per year and commercial demand by 0.1 percent per year. Unfortunately, energy savings from more efficient lighting and building upgrades are projected to be offset by increases in energy use elsewhere. For example, population growth coupled with in-migration to the Sunbelt increases air conditioning demands, and efficiencies gained from better household refrigerators and lights are offset by the increasing number of home electronics. Energy conservation alone will only slow demand, not decrease it.

The abundance of energy resources in the West ensures that the demand will be met there, at least in part. A recent U.S. inventory (Energy Policy and Conservation Act of 2008) shows the largest amount of future U.S. oil and gas resources coming from the West. Solar, wind, and geothermal resources are also likely to be concentrated in geographically distinct areas of the West, with wind in the high plains and solar in the desert Southwest.

Western states and provinces are already heavy energy producers. In 2007, the United States produced 76.5 exajoules domestically and imported the remaining 36.5 exajoules to meet demand (the balance, 5.5 exajoules, was exported). Canada produced 20 exajoules in 2006; nearly all the oil Canada produced but did not consume was exported to the United States (EIA 2008). Coal dominated U.S. production with 25 exajoules, 34 percent of which occurred in Wyoming and Montana, and nearly all Canadian coal was produced from large mines located in the western provinces (Stone 2007). Natural gas was the second largest source of energy produced in the United States (22 exajoules), 28 percent of which came from western states. Oil was the third largest source of U.S. energy production (11 exajoules), with 20 percent produced in western states (EIA 2009a). Nuclear energy produced 9 exajoules in the United States, with almost all active uranium mines in western states; Canadian uranium mining also occurs predominantly in western provinces. Renewable energy, including hydropower, made up the remaining U.S. energy production at 8 exajoules (EIA 2008). EIA (2009a) scenarios show renewable energy consumption growing at 3.3 percent per year for solar, biofuels, and wind, but fossil fuels remain the dominant energy source overall.

Americans' love affair with the West has created clusters of "40-acre ranchettes" around many western cities, carving up intact landscapes into low-density housing fragments. This exurban sprawl has become a primary environmental concern in the West. Increasingly apparent is a new threat of energy sprawl—the land area used for roads, wind turbines, wells, and transmission lines—that compounds the threat of exurban sprawl. With energy sprawl factored in, more than 206,000 square kilometers of land could be affected by new energy production by 2030 (McDonald et al. 2009). The increase in energy sprawl presents a "green dilemma" (chap. 8). All current sources of energy except nuclear have a large terrestrial footprint or carbon footprint. Siting new renewable energy sources in already disturbed habitats would decrease their footprint; along with decreases in air and water pollution, such measures make renewable energy more desirable for wildlife and for conservation as a whole.

Given the abundance of resources in the West and the species at immediate risk, this book covers energy resources (hydrocarbons, solar, wind, biofuels, geothermal, and nuclear) in the western United States and Canada likely to affect terrestrial systems. Hydropower is not covered because the impacts are largely aquatic and have already occurred. Offshore and onshore energy development in the East, Alaska, and the Yukon Territory is beyond the scope of this book.

To overcome the challenges of energy development in a place as socially valued and biologically rich as the West, we need a unifying vision for how to safeguard wildlife and allow development so that the right actions occur in the right places. To create that vision, the chapters that follow bring together the ideas of a diverse group of biologists, ecologists, and rangeland specialists representing a small nucleus of western federal and state agencies, nongovernment organizations, and universities that have been working on these issues. They have each pioneered and championed approaches to quantifying impacts of wildlife from energy development. Collectively, their studies show the similarities and challenges that species face with energy development and present a unifying vision and shared conservation strategies.

This story begins in chapter 2 with the likely extent and severity of future impacts in dominant biomes of the West. Chapter 2 uses the spatial tools of geographic information systems and the myriad publicly available datasets to provide an unbiased and holistic view of current and probable future energy development and the biomes affected. Although previous studies have shown that grasslands and shrublands are some of the least protected biomes in the West, chapter 2 highlights the immediate risk of oil extraction in the boreal forests of Canada. Everyone has a stake in the future of the West. The world expects the historical West to retain its wildness and wildlife, even if only a fraction of those people ever come to see it. The mere knowledge of its existence is a comfort. We need the West's oil, gas, wind, and other energy resources and yet also its essential character of wildness. Our choices will define this character well into the future.

Chapter 2

Geography of Energy Development in Western North America: Potential Impacts on Terrestrial Ecosystems

HOLLY E. COPELAND, AMY POCEWICZ,
AND JOSEPH M. KIESECKER

Rapid development of the rich energy resources found in western North America may have dramatic consequences for its vast areas with low human population density and undeveloped wild lands. If development continues at its current pace, the outcome will probably be energy sprawl (McDonald et al. 2009), resulting in a western landscape fragmented by energy infrastructure such as roads, well pads, wind towers, and transmission lines. Scientists increasingly warn of the threat posed by energy sprawl to iconic western species such as sage-grouse (*Centrocercus urophasianus*) and pronghorn (*Antilocapra americana*). Clearly, energy development is detrimental to many wildlife species, and the increasing demand for energy and the West's abundant supply nearly ensure that these resources will be developed. Our aim here is to illustrate the scale of potential impacts, to draw comparisons between different energy sources, and to catalyze large-scale planning efforts designed to meet energy demands while reducing impacts on sensitive wildlife species and habitats.

The energy demands of the United States are high, but so is domestic production. In 2008, energy consumption in the United States exceeded 104.5 exajoules (1 exajoule = 0.95 quadrillion British thermal units), with 78.1 exajoules produced domestically, and imports supplying the remainder of demand (34.8 exajoules). Canada consumed 14.8 exajoules and produced 20.4 exajoules in 2006; nearly all the oil Canada produced but did not consume was exported to the United States (Energy Information

Administration [EIA] 2008). Canada has large reserves of oil, natural gas, coal, and uranium, along with promising potential for development of wind and geothermal energy. Increasing political uncertainty in many oil-producing nations has prompted accelerating exploitation of North American energy resources, and growing recognition of the potential social and biological ramifications of climate change is driving trends toward increasing development of low-carbon or carbon-neutral energy sources such as solar, wind, nuclear, and geothermal power (Brooke 2008). If current trends continue, Canadian fossil and renewable energy resources probably will be developed rapidly and substantial proportions exported to the United States, with development limited mainly by the availability of transmission lines to carry energy from where it is produced to the highly populated areas where it is most needed.

Land tenure laws in the United States and Canada promote exploitation of energy resources. Lands managed by U.S. federal agencies such as the Bureau of Land Management (BLM) and the U.S. Forest Service make up roughly 43 percent of the western United States, and government-owned or Crown lands make up 60 percent of Alberta, 95 percent of British Columbia, 20 percent of southern Saskatchewan, and 95 percent of northern Saskatchewan. Overall, the Canadian and U.S. systems are similar in that prospective developers purchase a temporary right (called tenure in Canada) to develop or extract energy resources on government lands through competitive auction; however, in the United States, leases with no winning bids in a competitive auction are available as noncompetitive leases. Leases and licenses may be on lands where the government owns the surface rights or where the surface rights are privately owned and the minerals are government owned (called split-estate lands in the United States). Consequently, the areas of mineral rights retained by the government are often much larger than even the lands they manage. For example, the Alberta Crown holds the mineral rights to 81 percent of the province. In the United States, federal agencies are usually governed by multiple-use directives that have historically emphasized resource extraction (Knight and Bates 1995). Recently, the BLM has dramatically increased leasing for oil and gas development (Naugle et al. 2011) and has opened special renewable energy coordination offices to expedite development. Energy development on the 78 million hectares of land managed by the U.S. Forest Service has historically been limited, but recent controversial leases of oil and gas on national forest lands in places such as the Roan Plateau in Colorado and the Wyoming Range in Wyoming indicate growing pressure to develop energy resources wherever they occur.

Although demand for electrical energy in the United States has risen steadily since 1990, annual construction of new transmission facilities has not kept pace with demand (EIA 2009a). Efforts are under way to increase electrical transmission capacity and thereby reduce the primary technological constraint on new energy development in the West. New high-voltage (more than 230 kilovolts) transmission lines totaling 33,593 kilometers are proposed, which would augment capacity of the existing 99,875 kilometers of lines by roughly one third (Ventyx Energy 2009). The BLM is working on permitting projects in the western United States, such as the Gateway West and TransWest Express projects. Other proposed projects are international in scope, such as the Northern Lights project to link Alberta and Oregon. To increase efficiency of the growing number of transmission projects and address environmental concerns, the Western Governors' Association (2009) is supporting transmission corridor planning and facilitating coordination between the various participants, which would help stakeholder agencies move beyond the traditional project-by-project approach and toward proactive planning with a landscape vision (chap. 12, this volume).

Concerns about the environmental impacts of energy sprawl continue to draw the attention of scientists, policymakers, citizens, and environmental groups, yet the scope of the cumulative impacts on ecosystems remains largely unknown. Here we provide an overview of the major energy sources—both renewable and nonrenewable—with high potential for terrestrial impacts in western North America and quantify these impacts by terrestrial ecosystem. We provide an estimate of potential impacts from these major energy sources using lease and license data from the U.S. National Integrated Lands System database (BLM), Saskatchewan Mineral Disposition Maps and Databases, Alberta Energy, and British Columbia Ministry of Energy and Mines.

Major Energy Sources

We examine five major energy sources affecting terrestrial ecosystems in the West: hydrocarbons, nuclear, wind, solar, and geothermal. We do not consider hydropower because those impacts are largely aquatic, or the terrestrial impacts have already occurred. Nor do we consider biofuels, which are covered in chapter 8.

Hydrocarbons

The equatorial position of western North America during the Cenozoic and Carboniferous eras favored development of vast accumulations of petroleum (oil), natural gas, and coal, created when organic matter was deposited as sediments in basins. Supported by these abundant resources, production of conventional petroleum and natural gas has become, and is projected to remain, a dominant use of U.S. federal and Canadian Crown lands. Extraction of oil and natural gas typically creates a network of roads, pipelines, and well pads to access the resource, although newer technologies allow for directional and horizontal drilling that can dramatically reduce the surface footprint, and many companies have committed to significant onsite restoration projects. Active or pending oil and gas leases currently exist on nearly 59 million hectares of U.S. federal and Canadian Crown lands in western North America, with production occurring on leases occupying 25 million hectares (figs. 2.1 and 2.2).

Land use intensity is much higher for extracting conventional petroleum (oil), with a footprint of 369–2,114 thousand hectares/exajoule/year, than for extracting natural gas (150–880 thousand hectares/exajoule/year).

Deposits of oil sands (also known as tar sands) contain a combination of clay, sand, water, and bitumen. Oil sands are excavated in two ways: from open pit mining and from in situ techniques. With mining, open pits are created and the oil-rich bitumen near the surface is extracted, processed, then refined to oil (BLM 2009). With in situ techniques, the bitumen lies more than 75 meters from the surface and is extracted through wells in a process known as steam-assisted gravity drainage. Three deposits in northern Alberta, known as the Peace River, Athabasca, and Cold Lake oil sands, cover approximately 14 million hectares and are estimated to contain up to 309 billion barrels, including 173 billion barrels that are economically recoverable today (fig. 2.3).

There are 6.1 gigajoules (6.1×10^9) per barrel of oil, and the United States consumes 43 exajoules of liquid petroleum fuel annually (19.5 million barrels of oil per day), so 173 billion barrels of recoverable oil from the Athabasca oil sands would provide 1,055 exajoules of energy, or enough oil for approximately 25 years of U.S. demand. Because of its oil sands, Canada is second only to Saudi Arabia in terms of countries with the greatest oil reserves. Currently, 9.2 million hectares of Canadian Crown lands are leased for oil sand mining. Using data from Alberta Energy (2010) listing oil sand projects by barrels per day output, we calculated the average land use intensity for oil sands at 275,000 hectares per exajoule per year (fig. 2.2). Potential for oil sand mining in the United States is limited to deposits in eastern Utah covering 410,000 hectares, with an estimated 12–19 billion barrels. Extraction of these deposits is not commercially viable at present because

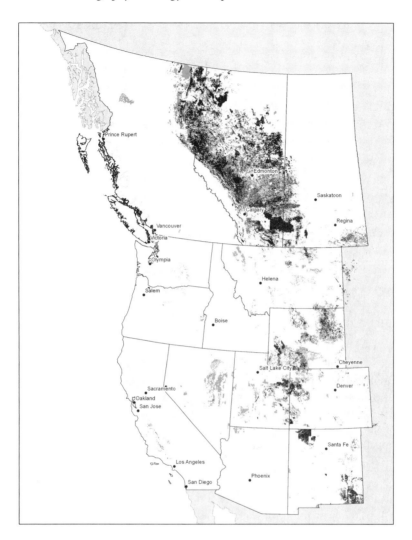

FIGURE 2.1. Oil and gas leases in the study area. Active and pending leases are shown in light gray, and the subset of these leases that are producing is shown in black. (Data on producing leases in the United States were acquired from the Bureau of Land Management National Integrated Lands System database, and those in Canada were derived from a spatial query in a geographic information system of lease and license parcels with producing oil and gas wells.)

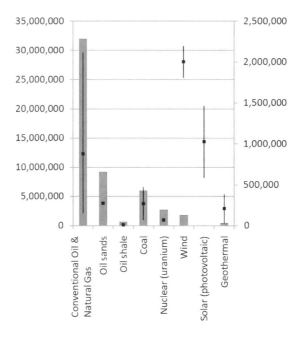

FIGURE 2.2. Area (hectares) leased (left axis) and land use intensity (hectares/exa-joule/year) of various energy sources from literature-based calculations and from McDonald et al. (2009; right axis). Oil sands and oil shale land use intensity numbers were derived using estimates of project area and barrels of oil per day. See chapter text for details. Conventional oil and natural gas data were combined so that the mean represents both oil and natural gas. Only producing leases are shown for conventional oil and natural gas.

the sands are wetted by hydrocarbon fluids, making extraction more difficult. Limited availability of water, in addition to substantial environmental concerns, further reduces economic viability of U.S. oil sand development.

Oil shale is a fine-grained sedimentary rock containing organic matter from which liquid hydrocarbons can be extracted; although no economically viable approach to extraction exists currently, research is under way. Vast oil shale deposits of the Green River Formation in Colorado, Wyoming, and Utah (BLM 2009) contain 1.5 trillion barrels of oil reserves, or five times Saudi Arabian reserves (Energy Information Administration 2009a). Oil shale can be extracted in one of two ways: through surface re-torting or mining and in situ retorting. Surface retorting involves an open pit mining process in which the rock is removed and processed. In situ re-torting involves heating the oil shale in place and extracting the liquid with

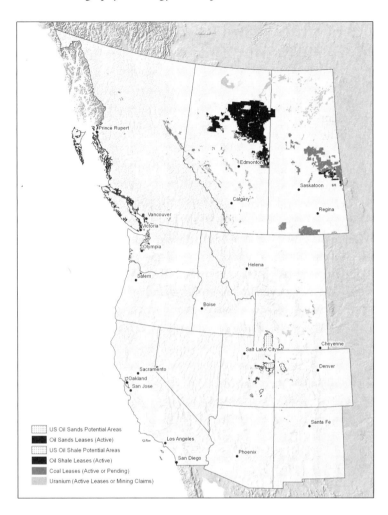

FIGURE 2.3. Map of oil sands, oil shale, coal, and uranium leases and potential in the study area.

little surface disturbance. Companies are experimenting with newer in situ retorting technologies that could make oil shale extraction viable within 25 years. Using the Piceance Basin example (Bartis et al. 2005), land use intensity calculations of oil shale development using in situ technologies are exceptionally small at only 37 hectares per exajoule per year (fig. 2.2). Oil shale leases in the region currently cover 679,000 hectares, but approximately 1.4 million hectares of lands in the Green River Formation contain oil shale deposits.

In addition to oil and gas, coal resources are abundant and mined throughout western North America (fig. 2.3). Of all energy produced in the United States, coal provides the largest share at 25.2 exajoules of energy annually (EIA 2008). The United States has the largest coal reserves in the world, and current production is second only to that of China. Canada ranks thirteenth in international coal production, with 1.6 exajoules produced annually. Canadian coal reserves represent greater energy potential than oil, gas, and oil sands combined; coal is currently the most important fossil fuel energy source in the country (International Energy Agency 2008). The twenty-nine large coal mines in the western United States produce 79 percent of the coal originating from mines in the United States, and 80 percent of this production comes from fourteen mines in Wyoming. Large coal mines also exist in Montana, Colorado, New Mexico, Utah, and Arizona. Nearly all of Canada's coal is produced from twenty large mines located in British Columbia, Alberta, and Saskatchewan (Stone 2007). Active coal leases currently cover approximately 6 million hectares in western North America. Coal mining, along with nuclear and geothermal power, affect small land areas relative to the amount of energy produced (fig. 2.2).

Nuclear

Uranium mining takes place throughout western North America; Canada is the world's largest producer and exporter (most is mined in Saskatchewan). Mined uranium supports 104 nuclear power plants in the United States and 7 in Canada. The uranium industry has experienced many booms and busts from price fluctuations. Price declines in 1980 effectively closed all active uranium mines, but rising prices in 2009 renewed interest in uranium mining and active uranium leases. The number of active uranium mines in the United States rose from four in 2003 to seventeen in 2008 (EIA 2009b), directly reflecting rising interest in nuclear energy and subsequent increases in the commodity price. Active and pending uranium leases covered 2.7 million hectares in 2009 (fig. 2.3). As society weighs the pros and cons of various sources of energy, interest in nuclear energy has renewed despite concerns voiced in the 1980s and 1990s over nuclear transport, production, and waste. Concerns about climate change have largely driven the new look at nuclear energy, given its low carbon footprint. In addition, even when the area of the plant, uranium mining, and waste storage are considered, nuclear power generation is a localized activity with a compact anthropogenic footprint compared with other energy sources (fig. 2.2).

Wind

The U.S. and Canadian projections suggest that wind resources may be able to provide 20 percent of annual electrical energy demand within the next 20 years. This would mean adding capacity for an additional 293,000 megawatts of wind power in the United States and 55,000 megawatts in Canada (U.S. Department of Energy 2008a; Canadian Wind Energy Association 2009). Large expanses of western North America have the potential for commercial wind energy development to help meet these goals (fig. 2.4).

Montana and Wyoming have the highest capacities for wind development, with a combined potential to generate more than 200,000 megawatts of power (American Wind Energy Association 2009). Commercially viable locations for generating wind power typically require minimum wind power densities of 300 watts per square meter and average wind speeds of 6.4 meters per second at a 50-meter height (Western Governors' Association 2009).

The 220 wind projects completed or under construction in the western United States have a combined generating capacity of 8,967 megawatts, yet they represent less than 3 percent of the total potential capacity of 339,070 megawatts present (American Wind Energy Association 2009). In western Canada, twenty-seven operating wind projects have an annual capacity of 702 megawatts (Canadian Wind Energy Association 2009). Many more projects are in the planning stages in both countries, and more than 1.8 million hectares of federal lands have authorized or pending leases for wind development in the United States or are affected by existing developments in Canada (fig. 2.4). At least 3,625 megawatts of additional wind power capacity is planned within the next several years across the three Canadian provinces. Wind power development is occurring on private lands, especially in the United States, but it is difficult to estimate the number of planned projects on private lands because there is no associated public leasing or environmental review process.

Wind power generation affects more land per unit of energy produced than hydrocarbon, nuclear, solar, or geothermal energy. The area occupied by wind turbines, access roads, and other infrastructure requires 2 million hectares per exajoule of energy generated (fig. 2.2), the highest land use

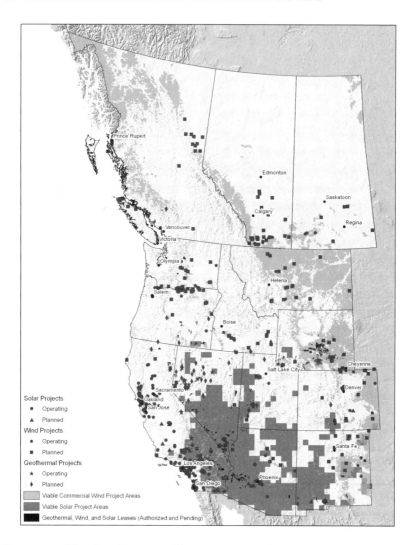

FIGURE 2.4. Map showing renewable energy areas and potential in the study area.

intensity of all energy sources studied. Thus, existing wind farms in western North America probably affect between 193,000 and 240,000 hectares of land. Recently the Western Governors' Association (2009) proposed renewable energy zones for the United States and Canada, which include both public and private lands, in all states and provinces except Saskatchewan. The areas designated as wind development zones encompass 14.3 million hectares. Although development would not be restricted to these zones, nor is development likely to affect the zones entirely, the zones do provide a coarse-scale estimate of the amount of land area that could be affected.

Solar

Generation of power from solar photovoltaic and solar thermal technologies more than doubled in the United States between 2000 and 2007. For these technologies to become more cost effective, 86,000–125,000 additional megawatts would have to be installed across the United States by 2030, with solar photovoltaic technology supplying 81 percent of that power (U.S. Department of Energy 2008b). Solar resources adequate for utility-scale projects are concentrated in the southwestern United States (fig. 2.4). The region currently has an installed capacity of 982 megawatts, 87 percent of it located in California. More than 97,000 hectares of federal lands have authorized or pending leases for development of solar energy, and areas identified as suitable for utility-scale solar development encompass 18.6 million hectares (Western Governors' Association 2009; fig. 2.4). Utility-scale projects consist of large solar arrays, and many of these projects are proposed in previously undeveloped landscapes. The area occupied by solar plants, access roads, and other infrastructure for photovoltaic power requires an average of 1 million hectares to produce an exajoule of energy per year; this is half the land use intensity of wind energy development but orders of magnitude greater than geothermal or nuclear energy (McDonald et al. 2009).

Geothermal

Geothermal power plants are currently operating in California, Idaho, Nevada, and Utah, producing 2,901 megawatts of power (U.S. Department of Energy 2008b). California has 60 percent and Nevada 15 percent of identified geothermal resources. Additional geothermal potential exists in British Columbia, Montana, New Mexico, Oregon, Washington, and Wyoming (U.S. Geological Survey 2008). Development of geothermal energy sources in the United States increased by only 5 percent between 2000 and 2007, but currently identified geothermal resources in the western United States could produce at least 3,355 megawatts. Furthermore, that level of energy production could increase 100-fold through use of enhanced systems consisting of engineered reservoirs that allow heat to be extracted from geothermal sources with low permeability or porosity (Massachusetts Institute of Technology 2006). More than 449,000 hectares of federal lands in the United States have authorized or pending leases for geothermal energy development, and more than thirty projects are currently planned (fig. 2.4). Geothermal energy has a much smaller footprint than other types of energy development, requiring an average of 208,333 hectares per exajoule per year of energy produced.

Cumulative Impacts of Energy Development on Terrestrial Ecosystems

The cumulative impact of current and potential future energy development on terrestrial ecosystems in western North America is largely unknown, but one estimate puts the cumulative footprint from new (i.e., 2009–2030) energy development in the United States as likely to exceed 20.6 million hectares (McDonald et al. 2009). Approximately 13 percent of the western United States is currently affected by an anthropogenic footprint (Leu et al. 2008), but similar estimates are not available for energy development alone or for Canada. Here we mapped the spatial distribution of all available pending and active energy leases across western North America, estimating the footprint of development on each of five terrestrial ecosystem types (temperate forests, boreal forests, shrublands, grasslands, and wetlands) (MEDIAS-France/Postel 2004; ESRI 2006). Wind lease data for Canada were unavailable, so we used existing projects to estimate minimum impacts (Ventyx Energy 2009). Each wind project point was expanded to represent the land area affected based on the power production of the project, assuming an impact of 20 hectares per megawatt (U.S. Department of Energy 2008a). We also calculated the footprint of proposed renewable energy zones for wind and solar energy development (Western Governors' Association 2009) to provide an upper estimate of the amount of land that may be affected.

Overall, we predict that new and existing energy development could affect, either directly or indirectly, up to 21 percent (96 million hectares) of the five major ecosystems in western North America, or 18 percent of all lands in the study area (table 2.1).

The highest overall predicted impacts as a percentage of the ecosystem type are to boreal forest (32 percent), shrublands (24 percent), and grasslands (21 percent). In absolute terms, the largest potential impacts are to

TABLE 2.1. Percentage and area (in hectares) of each terrestrial ecosystem potentially affected by energy development.

Type	Temperate Forest (152,273,205)	Boreal Forest (62,274,411)	Shrubland (169,341,030)	Grassland (76,913,703)	Wetland (239,778)	Total (percentage of all terrestrial ecosystems) (461,042,127)
Oil and gas leases	9.3 (14,212,734)	15.9 (9,911,502)	10.1 (17,158,285)	13.0 (9,971,273)	2.2 (5,238)	11.1 (51,259,032)
Currently producing oil and gas leases	4.1 (6,195,969)	6.4 (4,002,174)	3.6 (6,030,315)	5.8 (4,420,539)	0.6 (1,386)	4.5 (20,650,383)
Oil shale leases	<0.1 (127,548)	0.4 (257,418)	<0.1 (79,182)	0.2 (137,628)	0 (0)	0.13 (601,776)
Oil sands leases	0.4 (589,527)	10.6 (6,576,498)	0.5 (891,054)	1.1 (831,042)	<0.1 (18)	1.9 (8,888,139)
Coal leases	0.9 (1,373,301)	3.6 (2,266,965)	0.5 (773,721)	1.2 (954,981)	<0.1 (117)	1.2 (5,369,085)
Uranium leases	0.13 (196,578)	1.92 (1,198,305)	0.25 (423,261)	0.56 (432,099)	0.04 (90)	0.49 (2,250,333)
Wind leases	0.1 (209,016)	0 (0)	0.8 (1,338,921)	0.3 (200,736)	0.1 (342)	0 (1,749,015)
Wind zones (minus leases)	1.28 (1,942,731)	<0.1 (20,619)	3.31 (5,611,545)	3.76 (2,889,882)	1.38 (3,312)	2.27 (10,468,089)
Solar leases	0 (0)	0 (0)	<0.1 (59,868)	<0.1 (9,819)	0.1 (135)	<0.1 (69,822)

TABLE 2.1. Continued

Type	Temperate Forest (152,273,205)	Boreal Forest (62,274,411)	Shrubland (169,341,030)	Grassland (76,913,703)	Wetland (239,778)	Total (percentage of all terrestrial ecosystems) (461,042,127)
Solar zones (minus leases)	0.1 (137,250)	0 (0)	9.37 (15,867,162)	1.49 (1,148,310)	10.2 (24,480)	3.73 (17,177,202)
Geothermal leases	<0.1 (71,001)	0 (0)	0.2 (330,525)	<0.1 (23,832)	0.1 (315)	0.1 (425,673)
Hydrocarbon leases[a]	10.62 (16,176,021)	30.12 (18,754,966)	11.12 (18,833,428)	15.29 (11,757,845)	2.24 (5,373)	14.3 (66,118,032)
Renewable leases[b]	1.4 (2,150,982)	<0.1 (20,619)	12.88 (21,809,232)	5.28 (4,062,024)	11.72 (28,107)	6.09 (28,070,964)
All energy leases	12.25 (18,650,670)	32.49 (20,231,308)	24.29 (41,134,735)	21.31 (16,389,047)	14.00 (33,570)	20.92 (96,439,329)

[a]Oil and gas (all), oil shale, oil sands, and coal leases.
[b]Wind zones, solar zones, and geothermal leases.

shrublands (41,134,735 hectares). Boreal forests, shrublands, and grass-lands are especially vulnerable because of their geographic concurrence with the sedimentary basins that hold hydrocarbon deposits. These predictions probably inflate overall impacts because leased lands grossly overestimate areas that will ultimately be developed. A case in point is oil and gas development in the western United States, where significant portions of individual states such as Wyoming have been leased, but only small portions actually experience development. Conversely, impacts to some species are known to extend well beyond the actual development footprint (Doherty et al. 2011). Table 2.1, showing hectares of area leased and resulting impacts on ecosystems, should be interpreted with caution because these estimates are probably inflated. We use leases to provide an estimate of direct and indirect future impacts from energy development across our study area. Our findings are most useful for identifying broad trends and enabling comparisons of impacts between ecosystems.

The cumulative impacts of energy development from hydrocarbons probably will be greatest on boreal forests (30 percent) and grasslands (15 percent). Among types of hydrocarbon development, conventional oil and gas have the most land area leased for current or future production (fig. 2.1). Oil sands and coal mining could affect a disproportionately large percentage of boreal forests in Canada (11 percent and 4 percent, respectively). Potential impacts to other ecosystems are smaller, though not insignificant (table 2.1). A limited number of lands are currently leased for oil shale, with minimal (less than 1 percent) predicted impacts to terrestrial ecosystems, suggesting that although the long-term potential for significant development exists, current leases for oil shale development are largely experimental.

The overall impact of uranium mining is likely to be small, potentially affecting 0.6 percent of all terrestrial ecosystems we studied. The largest predicted impacts are to the boreal forests (2 percent) in Canada, followed by grasslands (0.6 percent) located mostly in Wyoming (fig. 2.3).

Wind development will probably have the greatest impacts on shrub-lands at 0.8–3.0 percent and grasslands at 0.3–4.0 percent (table 2.1), ecosystems where many wildlife species are negatively influenced by the presence of tall structures. Current wind leases represent only about 11 percent of potential in a maximum-development scenario (Western Governors' Association 2009). Solar leases currently occupy a small area, exclusively on nonforested ecosystems, but these leases represent only about 3 percent of areas with potential for development in Western Governors' Association zones. Therefore, current wind and solar leases probably greatly underestimate the future impact on terrestrial ecosystems. Geothermal leases also occupy a small area but are most concentrated in shrublands (0.2 percent), like many of the other energy types.

Overall, conventional and unconventional oil and gas development have and probably will continue to have a large footprint, especially on the boreal forests in Canada but also on grasslands, shrublands, and temperate forests. However, if renewable energy development proceeds on a maximum-development basis, then new solar and wind power generation could significantly affect shrubland and grassland ecosystems in the United States.

Conclusion

The changes predicted to shrubland and grassland systems in the United States and boreal forests in Canada are especially disconcerting because these systems currently receive little legislative protection, yet they support key inhabitants such as sage-grouse, pygmy rabbit (*Brachylagus idahoensis*), boreal woodland caribou (*Rangifer tarandus caribou*), and Wyoming pocket gopher (*Thomomys clusius*) that are being considered for protection under the Endangered Species Act. In addition to impacts associated with energy development, these systems are also suffering under stresses from residential development, invasive species, disease, and climate change.

The history of the West has been marked by prolonged confrontations between resource use and conservation, and the distribution of energy resources in the region ensures that such conflicts will continue and probably intensify. Policymakers face the complex task of balancing competing environmental and energy independence concerns, various federal agency interests and missions, and the numerous permutations of available energy resources. The goals of energy development and conservation need not be mutually exclusive, but reducing environmental damage will entail a fundamental transformation in how we think about planned development (chap. 9). Understanding the scale of anticipated impacts on Western ecosystems may motivate policymakers to engage in proactive planning, ideally before projects begin, about how to avoid siting conflicts, maintain biodiversity, and determine suitable mitigation responses. Reducing impacts on wildlife also will entail much greater investment in offsets (compensating conservation actions) to address residual project impacts and deliver net gains for nature (Kiesecker et al. 2009, 2010).

PART II

Biological Response of Wildlife and Invasive Plants to Energy Development

Science provides the biological basis for formulating conservation actions and shaping policies that permit sustainable development while safeguarding wildlife. Paramount to linking science with management is the ability to ask the right questions in a way that is relevant to decision makers. Equally important for producing credible science is scrutiny of our research approach to ensure that its rigor stands up to the highest levels of peer review. The best and most recent studies are those that evaluate cumulative effects, defined here as the synergistic, interactive, and sometimes unpredictable outcomes of multiple land use practices, including energy development, that aggregate over time across broad landscapes.

Early studies typically evaluated short-term and behavioral avoidance by an individual animal of the drilling of one or a few exploratory energy wells. Such studies provided the impetus for later research but alone are inadequate to evaluate cumulative effects of developments large enough to influence the hundreds or thousands of animals that make up wildlife populations. More obvious in recently published literature are well-designed studies that evaluate cumulative effects of development at the scale of the population rather than the individual animal. These improvements provide decision makers with the ability to reliably predict the outcomes of alternative management scenarios to conserve at-risk species. The new yet largely unexplored frontier in science extends beyond the species of interest to include possible effects of energy development on interactions between species that may drastically alter predator–prey dynamics and other inter-specific relationships.

Society places a premium on science and its role in natural resource management. The most important thing to keep sight of is our desire to produce the most credible science possible. Readers will see firsthand that most wildlife–energy impact research is in its infancy, that much has been learned in the last 5–10 years, and that many unanswered questions remain. In total, we know enough to move forward in conservation with the under-standing that our knowledge will never be perfect. Historically, available funding supported studies on big game and other charismatic species. But as the public interest in biodiversity conservation grows, so does our result-ing knowledge for an array of other plant and animal life.

A sampling of the recent science reveals the wealth of new information available and the ways in which we can improve the rigor and depth of the resulting science. Part II represents the first time that recent wildlife impact studies have been synthesized and made widely available. Chapter 3 pres-ents a unifying approach for the design and interpretation of research to evaluate potential effects of human developments on wildlife. This ap-proach provides scientists with a framework for designing new studies and gives decision makers a basis for evaluating the merits of existing studies for use in policy and management decisions. Chapter 4 synthesizes the impacts on sage-grouse (*Centrocercus urophasianus*) of oil and gas development and lays out a conservation strategy for maintaining a set of large, intact popula-tions. Recent research also documents thresholds of impact relative to de-velopment intensity that can be used to forecast biological tradeoffs of newly proposed drilling. When additional drilling is approved, the authors show how thresholds can be used to offset impacts by conserving an equal or greater number of birds in priority landscapes.

Chapter 5 provides a mechanistic understanding of population-level declines in woodland caribou (*Rangifer tarandus tarandus*) attributable to increased predation by wolves (*Canis lupus*) along energy roads and seismic lines cut through the Canadian boreal forest. Management models that suc-cessfully predict caribou population growth rates as a function of the per-centage energy development and percentage area burned confirm the grim predictions of the cumulative effects of landscape change. The newest re-search has worked out the economics of conservation triage for caribou in hopes of proactively applying conservation over a broad enough landscape to save a few representative populations. Chapter 6 synthesizes studies showing that ovenbirds (*Seiurus aurocapilla*) do not incorporate energy seismic lines into their breeding territories. A follow-up control impact design with and without seismic lines shows declines in the number of territories with increasing seismic line density, suggesting that behavioral avoidance of lines may lead to local reductions in songbird abundance. Chapter 7 is a call for research to elucidate the role of energy infrastructure as a conduit for invasive plants. The authors provide a compelling case for decision makers to consider the risks of biological invasions in landscapes where human disturbance in the form of energy development alters critical ecosystem processes. Chapter 8 shows that with anticipated increases in renewable energy, careful planning is needed to avoid conflicts between development of green energy and concerns about wildlife impacts. If such developments are properly sited and sufficiently mitigated when impacts are unavoidable, we can finally advance renewable energy production beyond current levels without compromising wildlife conservation.

Chapter 3

Unifying Framework for Understanding Impacts of Human Developments on Wildlife

CHRIS J. JOHNSON AND
MARTIN-HUGUES ST-LAURENT

Natural resource professionals recognize the negative impacts of human developments on the distribution, abundance, and, in some cases, persistence of wildlife populations or species. Indeed, human activity in all its forms (Kerr and Currie 1995) is a primary cause of the global decline in biodiversity in general (Brooks et al. 2002; Dudgeon et al. 2006; White and Kerr 2006) and wildlife in particular (Ceballos and Ehrlich 2002; Laliberte and Ripple 2004; Davies et al. 2006). This recognition has led to a rapid increase in the number of studies designed to elucidate and document wildlife–human interactions (fig. 3.1).

Ranging from site-specific wildlife–human encounters during recreational activities (e.g., Naylor et al. 2009) to the large-scale development of oil and gas reserves (e.g., Bradshaw et al. 1997), the effects can differ greatly; however, many of the types of impacts are comparable. We use the term *effect* to mean a change in the environment resulting from a human activity and the term *impact* to represent the consequences of such changes for wildlife populations (Wärnbäck and Hilding-Rydevik 2009). This distinction is important because all developments will have some effect, but the impacts will vary according to a range of factors.

The distribution of wildlife populations and the occurrence of natural resources interact with past land uses and regulatory frameworks to impact the ecology or biology of a species. Impacts can include changes in animal behavior, energetics and nutrition, physiology, distribution, resource use,

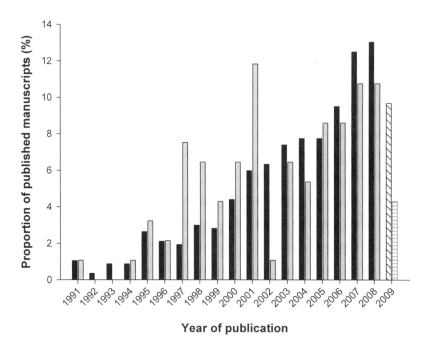

FIGURE 3.1. Number of scientific publications assessing human impacts (black bars) and cumulative effects (gray bars) on wildlife species since 1991. Frequency of publications was tabulated using ISI Web of Science and the search terms "human impacts and wildlife" and "cumulative effects and wildlife." Dashed bars in 2009 refer to incomplete years of data, as publication was still ongoing at the time of the search.

population dynamics, and interspecific interactions leading to changes in plant and animal communities. The dynamics of effects vary both spatially and temporally (box 3.1).

At the coarsest scale, the types of development and associated effects are a function of the geographic distribution of natural resources. Where resource development opportunities occur, some jurisdictions are more likely to encourage a high rate of resource development as a means to support or grow an economy, whereas other jurisdictions may have more restrictive environmental regulations that emphasize other values (Timoney and Lee 2001; British Columbia Oil and Gas Commission 2004). Finally, societal expectations, regulatory frameworks, and even the economic viability of natural resource sectors may change over time (Jackson and Curry 2002).

BOX 3.1. DYNAMIC IMPACT OF DEVELOPMENT ON WILDLIFE IN NORTH AMERICA

Before colonization of North America, indigenous peoples used wildlife and influenced the environment, but the magnitude of these impacts is debated and
probably variable (Martin and Szuter 1999; Sherry and Myers 2002; Laliberte and
Ripple 2003). After settlement by Europeans, agriculture and unsustainable harvest had major influences on wildlife populations and their habitats (Mattson
and Merrill 2002; With et al. 2008; Brown and Boutin 2009). Through the twentieth century the range of activities increased and the area and magnitude of effects grew to encompass most areas of the continent. Management and conservation agencies have made progress in setting and enforcing harvest to
sustainable levels, and in some areas agricultural lands are being reclaimed by
the forest, with benefits for wildlife (Lancaster et al. 2008).

However, new threats have emerged to replace old challenges. Exploration
and development of oil and gas, mining, and forestry, at an industrial scale, result in isolated site-specific and cumulative impacts for wildlife populations, often over very large areas (Berger 2004; Johnson et al. 2005; Vors et al. 2007;
Nitschke 2008). In the second half of the twentieth century and into the
twenty-first century the newest and largest challenge for regulatory agencies
and biologists is energy development. North Americans' reliance on fossil fuels
has resulted in widespread exploration and development of conventional
sources of oil and natural gas. Nonconventional sources of hydrocarbons, such
as coal bed methane and bitumen deposits, are now economically viable but
have a much larger ecological footprint (Johnson and Miyanishi 2008). Even
new green sources of energy, such as wind and solar, can result in measurable
environmental impacts for wildlife (Kuvlesky et al. 2007; Pruett et al. 2009b;
but see Devereux et al. 2008; Sovacool 2009).

In response to the societal challenge of balancing current levels of economic development and ecological values, wildlife and conservation ecologists have focused on quantifying impacts and understanding the implications of industrial development. Although researchers and practitioners
have made much progress in the past 25 years (fig. 3.1), we have found little evidence of consensus or even discussion of unifying theory, acceptable
definitions, or general methodological frameworks for addressing these important questions. This lack of dialogue is probably a product of perceived
taxonomic, life history, or effect-specific differences between projects designed to reveal the impacts of specific wildlife–development interactions.
However, a lack of unity adds to confusion when impacts are interpreted,
especially when statistical effect sizes are small.

We recognize the challenges in finding a unifying theory of impact research and interpretation. Ecology as a science has long struggled to define general theory and principles that apply across space, time, and taxonomy (Pigliucci 2002). In the context of impact science, however, we believe that improvement is possible and needed. Much progress can be made by explicitly recognizing the pathway and relative magnitude of effects, the spatiotemporal dimensions of the effect, the regulation and mitigation of effects, and the biological scale of the impact for the organism. In combination, these four elements can serve as a unifying framework to categorize and predict the potential magnitude of impacts of energy development for a wildlife population; to relate the effects of development to the realized impacts for a population, an important component of the regulatory and mitigation process; and to compare the range of expected or observed impacts across populations using a consistent set of measures and definitions.

In this chapter, we draw on the four elements to construct a generic typology for wildlife impact research. This typology is premised on the principles of scale, a concept that is now widely appreciated by ecologists and has utility for directing research design and conveying findings to managers, policymakers, and legislators. Although the focus of the supporting chapters is on energy development, we take a broader perspective, reviewing the pertinent literature from across the applied ecological, management, and conservation sciences that addresses the full range of wildlife–human interactions. Our objective is not to provide a formula but instead define a general framework of ideas, principles, and methods that practitioners and scientists can call on when making comparisons across spatial, temporal, population, and taxonomic boundaries.

A Typology for Wildlife Impact Research: The Framework

Responses of wildlife to effects of human development are complex, variable, and scale-specific (Merrill et al. 1994; Gucinski et al. 2001; Blondel 2008; Wallgren et al. 2009). Furthermore, the metric we choose to define and measure an impact, as well as regulatory or mitigation measures, will influence the significance of any finding. In an effort to identify and categorize this complexity, we propose three broad categories of effects that capture the causal mechanisms regulating the magnitude of impacts caused by human development (fig. 3.2).

Evidence suggests that wildlife populations will suffer a greater degree of impact as the spatiotemporal scale of effects increases; the pathway of the

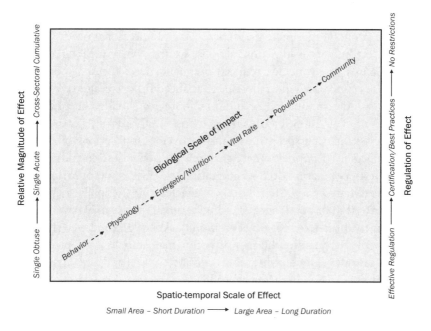

FIGURE 3.2. Schematic representation of a typology for classifying and predicting the impacts of human–wildlife interactions.

effect changes in a way that results in more severe, longer-lasting, and less predictable impacts; and the relative magnitude of the effect increases as a product of ineffective regulatory or management processes.

Relating effects to impacts, we have conceived of a range of responses of wildlife populations that vary hierarchically, probably in a nonlinear way, as the effects increase in severity (fig. 3.2). Although we do not suggest a simple incremental and discrete relationship, the scale of biological impacts is represented well by the techniques biologists use to elucidate the interactions between resource development activities and wildlife. At the scale of the individual, animals in the vicinity of human activities can demonstrate a change in behavior, such as movement away from a disturbance, reduced hunting efficiency, or altered social dynamics (Galanti et al. 2006; Rabin et al. 2006; Naylor et al. 2009). Although species- and disturbance-specific, this level of response could be followed by a physiological reaction, such as increased heart rate (MacArthur et al. 1979; Krausman et al. 2004; Thiel et al. 2008; Bisson et al. 2009). A more severe but not independent impact is an energetic or nutritional expense resulting from increased movement, reduced caloric intake, or failure to access important and rare resources (Tyler

1991; Bradshaw et al. 1998; Reimers et al. 2003). If an animal or group of animals can no longer meet their energetic and nutritional needs or suffer from direct human–wildlife interactions, we may observe changes in vital rates. Causes could include reduced pregnancy rates through poor body condition or increased mortality due to reduced fitness or altered behavior, such as risk-prone foraging or hunting strategies (Phillips and Alldredge 2000; Frair et al. 2008; Fahrig and Rytwinski 2009). From a conservation and management perspective, the most extreme implications of development effects for a population are changes in distribution, including range contraction, population decline, and ultimately extirpation or extinction of a species (Mattson and Merrill 2002; Schaefer 2003). Finally, a change in the distribution or abundance of a single animal species may have implications for the broader ecological community. Although food web dynamics are difficult to quantify, human activities may initiate or contribute to trophic cascades (Hebblewhite et al. 2005b; Borrvall and Ebenman 2006; Anthony et al. 2008; Berger et al. 2008).

We see the range of effects and impacts as proceeding across a continuum of space, time, and regulatory response. Thus, scale provides a functional linkage, although in a nonconventional form. We begin with a more detailed explanation of scale, from an ecological impact perspective, and then provide an overview of the three sources of effects. We then review the impacts literature to illustrate the biological scale of impacts presented in figure 3.2. Finally, we conclude with a brief overview of issues and practices that may help guide study design in the context of wildlife impact research.

Scale: A Unifying Principle

Most biologists now appreciate the importance of spatial and temporal scale when designing and interpreting ecological studies (Wheatley and Johnson 2009). Although the idea of spatiotemporal variation in ecological process and observation has a long tradition in the ecological literature, only recently has scale received explicit consideration from methodological and theoretical perspectives (Schneider 2001). The now prominent role of scale as a primary element of research design can be traced back to the mid-1980s. At that time, researchers in a number of applied ecological subdisciplines began to note a general lack of appreciation for cross-scale phenomena and the associated risks to inference (Senft et al. 1987; Wiens 1989; Kotliar and Wiens 1990; Levin 1992). Through the 1990s the earlier and mostly speculative writings of scale were put into practice. Researchers designed studies to measure variation in animal responses across arbitrary breaks in spatiotemporal scale and later to identify holons or boundaries based on behavior or activity (Schaefer and Messier 1995; Wallace et al. 1995; Saab 1999; Johnson et al. 2002; Nams and Bourgeois 2004). Likewise, our conceptual definition of scale has evolved over time. Scale was initially considered as something that bounded our sampling or inferential framework. Drawing on one of the early works, Turner et al. (1989: 246) defined scale as the "temporal or spatial dimensions of an object defined by both grain and extent." Building on that initial premise, the recognition of scale as a static element of research design has expanded to consider scale as a dynamic principle that captures the range of behaviors of animals as they respond to the environment. Thus, holons in scale are defined not by precise sampling thresholds that correspond with time or space but by distinct changes in behavior, perhaps as a function of time and space. In an effort to advance our application of scale-related principles and reduce confusion resulting from myriad definitions, Dungan et al. (2002) identified three broad categories of scale: observation, analysis, and phenomenon. This later category, the spatiotemporal dimensions of the object being studied and the processes that affect it, is of greatest relevance to understanding the relationship between human developments and wildlife impacts.

Our understanding of the importance of spatial, temporal, and ecological scale for impact research has improved greatly over the past 30 years. For one of the best-studied species, caribou (*Rangifer tarandus*), research on the impacts of human activity and infrastructure development has shifted from small-scale behavioral research to larger-scale landscape studies focused on population distribution and abundance (Vistnes and Nellemann 2008). This trend toward quantification of population-level impacts across regional areas is consistent for other species (Schneider et al. 2003; Johnson et al. 2005; Nielsen et al. 2008). With these conceptual and methodological advances our understanding of wildlife impacts has improved, and that knowledge is playing an increased role in decision-making and monitoring frameworks (Hood and Parker 2001; Johnson and Boyce 2004; Diavik Diamond Mines Incorporated 2008; Bennett et al. 2009). However, opportunities remain for progress in how we conduct multiscale studies and how they inform our understanding of the impacts of human developments (Wheatley and Johnson 2009). As a starting point, we suggest that the principle of scale can serve as a general integrative concept for assessing the magnitude of effects and characterizing wildlife impact studies.

Spatiotemporal Scale of Effect

The ecological scale of animal responses has direct links with regulation. Environmental assessment studies designed to facilitate the regulatory approval and mitigation processes often identify a spatiotemporal zone of influence. This is the area and perhaps time of year when wildlife experience significant impacts, such as an avoidance response, where animals shift their distribution away from a development, alter behavior in the vicinity of a facility, or change the types or quality of habitats used. Although it is an intuitive concept, the zone of influence and measures of significance are difficult to quantify (Quinonez-Pinon et al. 2007). This is especially apparent where multiple developments interact in a cumulative way. Also, the zone of influence should be premised on the type of response observed; there may be multiple zones depending on the source of effect. Data also suggest that the zone of influence varies according to the development type. Johnson et al. (2005) documented an avoidance response near diamond mines for barren-ground caribou (*R. t. groenlandicus*) that exceeded 100 kilometers; human activities with smaller and less intensive footprints were also avoided but over a smaller area. Other researchers reported avoidance behaviors and corresponding zones of influence of caribou at finer spatial and temporal scales. For example, Nellemann and Cameron (1996, 1998) found that the density of barren-ground caribou (*R. t. granti*) was inversely related to road density and that caribou avoided high-quality habitats within 4 kilometers of roads and oil production facilities. Woodland caribou (*R. t. caribou*) in northern Alberta demonstrated an avoidance distance of 1,000 meters for oil and gas wells and 250 meters for seismic lines, accounting for a 22–48 percent decrease in habitat availability (Dyer et al. 2001). Recent research has focused on developing techniques that indicate statistically meaningful responses of animals to human activities or facilities that can then be translated to zones of influence used in regulatory frameworks (Bennett et al. 2009). When empirical data are absent or the review process requires an inadequate level of scientific rigor, expert opinion is used to estimate probable zones (e.g., AXYS and Penner 1998). Often, the processes used to collect such ecological data are flawed (Johnson and Gillingham 2004).

Considering more traditional definitions of scale (e.g., Turner et al. 1989), the spatial or temporal dimensions of an effect resulting from a human development often correlate positively with the magnitude of impacts for wildlife populations. In general, as the zone of influence of a development increases over time and space we observe a greater likelihood of inter-

BOX 3.2. RESILIENCE: A UNIFYING FRAMEWORK FOR IMPACT RESEARCH

Resilience is the capacity of an ecological system to absorb disturbance or environmental variation before transitioning to an alternate stable state (Holling 1973). Typically, resilience reflects a complex set of interrelated and dynamic relationships that occur across a panarchy (Gunderson and Holling 2002). As theorized, a system might achieve an alternate state if the threshold level of some controlling variable is passed, leading to a feedback mechanism, which will redirect the trajectory of the system. In the context of wildlife, feedbacks resulting from human activities could lead to a population increase or decrease, or a change in distribution not associated with some long- or short-term cyclical pattern (e.g., seasonal change in range). At the extreme of this continuum, a new stable state could be a population that is forced to a historically low density or even the extinction of the species. Resilience theory is now accepted in management and conservation arenas as a unifying framework forcing linkages between patterns, mechanisms, variation in key environmental factors, and human activities (Folke et al. 2004).

actions with a larger proportion of a population, for a larger number of species, and fewer opportunities for those populations to find alternative habitats. These scale-related dynamics are represented in the typology as the spatiotemporal scale of effect (fig 3.2) and are captured by the broader principle of ecological resilience (box 3.2).

Numerous examples illustrate the incremental responses of wildlife populations to the increasing spatiotemporal magnitude of human activities. Survival of grizzly bears across most of North America is a function of the frequency of human–bear interactions and associated management responses (Pease and Mattson 1999; Nielsen et al. 2004; Schwartz et al. 2006). When considered over large spatiotemporal scales, human-caused mortality has resulted in the near extirpation of grizzly bears from the United States (Mattson and Merrill 2002; Pyare et al. 2004). Schaefer (2003) reported that nearly half of the historic range of woodland caribou was lost over a 100-year period in Ontario, Canada (34,000 square kilometers per decade), a trend that correlated with forest development and other human activities. As with grizzly bears, a combination of effects that are continuous over a long period of time and broad geographic area resulted in severe population-level impacts. Although such studies provide compelling evidence, the relationships between long-term or broad-scale human activities and wildlife impacts are not always linear in time or space.

Vors et al. (2007) reported that the extirpation of caribou across Ontario was most strongly related to forest harvesting, when compared with other human activities, but only after a two-decade lag and within a tolerance threshold of 13 kilometers to the nearest clearcut. In contrast, mule deer (*Odocoileus hemionus*) in Wyoming avoided gas well pads within the year of construction (Sawyer et al. 2006). The dynamics of barren-ground caribou provide an excellent example of the panarchy that characterizes human impacts in the context of natural population variation and resilience. For one herd, traditional ecological knowledge and dendrochronological evidence have revealed an approximate 40-year naturally occurring population cycle (Zalatan et al. 2006). Working in the context of population cycles, biologists must assess the recent impacts of resource development. Disturbance or habitat change could increase the period or amplitude of the cycle, with broad implications for dependent aboriginal communities. In the context of barren-ground caribou, some claim that decoupling natural from human impacts is nearly impossible (Cronin et al. 1998). Although difficult, this is a challenge that must be confronted if we are to improve the strength of inference of wildlife–human impact studies.

Magnitude of Effect

The effects of a development can manifest as impacts with increasing severity and implications for population persistence and community dynamics. This can be a direct result of the total area and time of exposure to the effect, but also the type and severity of development activity (fig. 3.2). We use the term "magnitude of effect" to represent this relationship. Although researchers place much emphasis on negative human–wildlife interactions, not all hypothesized impacts are confirmed by data. Following comprehensive studies, Krausman et al. (2004) reported only a small response in the behavior of Sonoran pronghorn (*Antilocapra americana sonoriensis*) to military activity, Devereux et al. (2008) found no impact of wind turbines on the distribution of farmland birds, and Tyler (1991) detected minimal behavioral and energetic responses of Svalbard reindeer (*R. t. platyrhynchus*) to direct provocation by snowmobiles. In these cases and others, a number of interacting factors probably regulate the total effect and resulting impacts (fig. 3.2), but a small observed impact also may simply reflect a small or trivial effect. Thus, estimating and reporting the absolute or relative magnitude of effect is essential when conducting wildlife impact studies.

The magnitude of effect spans a range of wildlife–development interactions that includes single activities with an obtuse outcome for the environment, followed by acute but isolated effects, to the extreme of large-scale acute effects that are cumulative over time and a range of development activities. An obtuse or imperceptible effect is characterized by a low frequency of occurrence, a small spatial footprint, or a limited sensory experience for wildlife. In the context of energy development, a single cutover for seismic exploration of oil and gas in forested areas probably has a very limited long-term impact for ungulates; the width of the seismic line is small (1–8 meters) relative to other disturbance types, geophysical exploration activities have temporally limited acute effects (Bradshaw et al. 1998), and the presence of humans is infrequent on an annual scale (Lee and Boutin 2006). Progressing along the effect continuum, a network of seismic lines may result in a large area of habitat change, greater use by humans, and a change in the functional response of predators (James and Stuart-Smith 2000; Dyer et al. 2001; Schneider et al. 2003; McCutchen 2007).

Characterizing the effect space is challenging and depends on the species of interest, the disturbance type, the metric used to assess the impact, and the spatiotemporal scale of observation. For example, individual reindeer may show only a small energetic response to the immediate provocation by snowmobiles (Tyler 1991). However, at a larger scale using a different impact metric, woodland caribou have abandoned significant portions of winter range after repeated disturbance by the same effect source: snowmobiles (Seip et al. 2007). In some situations, the effect may be consistent, but the measured impact varies according to research design and interpretation (Cronin et al. 2000; Cameron et al. 2005).

Continual development or increasing human populations may have cumulative effects across regions. These incremental effects result in an increase in the severity or unpredictability of impacts and are the most extreme outcome for wildlife relative to the range of pathways we identified (fig. 3.2). We define cumulative effects as the synergistic, interactive, or unpredictable outcomes of multiple land use practices or development that aggregate over time and space (Ross 1998; Harriman and Noble 2008). Among the types of industrial development we find across North America, the energy sector is prone to cumulative effects. Exploration and development for conventional and nonconventional sources of oil and gas use large areas of land with overlapping tenures for agriculture, forestry, and mining. A recent study for north-central British Columbia, Canada, measured the cumulative effects of 35 years of forestry, oil, and gas and agricultural development and found both additive and synergistic impacts related to a reduction and fragmentation of habitat for a number of species (Nitschke 2008). Similarly, Sorensen et al. (2008) developed an empirical relationship between the discrete growth rate of populations of boreal woodland caribou in Alberta and two measures of cumulative effects. The regression was an excellent fit and serves as a tool to evaluate further industrial development in the context of population targets for caribou.

Across parts of North America, resource development, including carbon-based and non–carbon-based energy, is in the exploration and planning phase. Thus, opportunities still exist to address cumulative effects before long-term tenures and leases are issued, and the economic imperative limits the range of actions we might consider for minimizing or reducing impacts over regional landscapes. This is especially true for the boreal and tundra regions of northern Canada and portions of the western United States (chap. 2). Fortunately, practitioners and researchers are recognizing this opportunity for proactive solutions, with much emphasis on the development and application of techniques for quantifying cumulative impacts on wildlife populations. These efforts are supported by better resource inventories, including data describing the distribution of wildlife species. Also, growth in the power and applicability of species distribution, simulation, and disturbance models has strengthened research on cumulative effects (Schneider et al. 2003; Johnson et al. 2005).

Regulation and Mitigation of Effect

Biologists often focus their efforts on understanding and communicating the level of impacts. However, we suggest that the outcome of human developments for wildlife populations is most directly related to the types and effectiveness of regulatory frameworks (chap. 11). From the perspective of industrial development, legislation addressing environmental impacts is the most pertinent (Johnson 2011). In Canada, federal and provincial assessment processes regulate a defined set of developments. Each set of legislation has a specific set of trigger mechanisms; effects that occur outside large energy development projects may be considered by other environmental law and policy processes. For example, across a special management area of British Columbia, Canada, pretenure plans are required before companies receive leases to extract oil and gas. Across that same area, best practice guidelines direct exploration activities so that they minimize the impacts for environmental, cultural, and recreational values (British Columbia Oil and

Gas Commission 2004). Comprehensive legislation governing forest practices in the province enables the review and approval of harvesting plans
and provides mechanisms to protect specific wildlife habitats and representative tracts of old-growth forests (Johnson 2011). This is just one of many
possible examples of overlapping regulatory frameworks. Policy and legislative tools vary between countries, states, and provinces. Also, the administrative or political will to implement, enforce, and adapt such tools to
evolving or intensifying resource sectors may vary, especially when local or
national economies depend on stable or increasing levels of extraction
(Jackson and Curry 2002).

As an alternative to government-initiated policies and legislation, many
corporations promote self-regulated sustainable practices and a corporate
ethic that is responsive to the needs of local communities and the health of
the environment (Auld et al. 2008a). In some sectors, nonlegislated certification processes are widely adopted as formal mechanisms to prove compliance with broadly accepted principles and criteria for sustainable practices.
Certification systems are most well developed and adopted for the forestry
sector (Auld et al. 2008b). Across North America, many companies now
adhere to one or more certification schemes, with significant positive implications for nonregulated practices and policy (Cubbage and Newman
2006; Fraser 2007). Such market-based mechanisms are especially useful
when the consumer is educated and discerning.

The level of effect and resulting impacts for wildlife will be a product of
legislated or nonlegislated restriction or guidance for industrial developments. This can include simple project approval or rejection but often involves a process in which approval is contingent on a set of practices or mitigation strategies that minimize environmental effects. Depending on the
rigor of these processes and the will of government and corporations to ensure effective environmental restrictions, the range of impacts for wildlife
populations could vary widely (Mooers et al. 2007; Findlay et al. 2009).
Effectiveness of regulation and policy, including compliance and enforcement, is a complex subject based in politics, economics, culture, and psy-
chology (Jones et al. 2008; Keane et al. 2008; You 2008; Gaveau et al.
2009). Although specific to each jurisdiction, we suggest that impacts are
most limited or mitigated where effective legislation is in place and enforced (Rowcliffe et al. 2004). Thus, according to the typology, the large-
scale cumulative effects and corresponding impacts at the population or
community level could be reduced by effective regulatory processes that restrict activities, reject components of a proposal, or demand the redesign of
certain elements of a project (Sinclair 2000; Zellmer 2000). In some cases,

mitigation of effects or reclamation activities might also substantially reduce short- and long-term impacts. We argue that impacts become more
likely or unconstrained as industry is allowed to self-regulate through nonlegislated certification or best practice approaches independent of effective
legislation (Kimerling 2001; Ebeling and Yasue 2009). This should not be
construed as a blanket criticism of such approaches (Newsom et al. 2006);
such methods may be the most effective tool for constraining environmental effects in some situations (Melo and Wolf 2005). We suggest that the
greatest level of impact occurs where government or industry action is ineffective or lacking (fig. 3.2). In these cases, industrial development is focused on production of commodities, with no limits in respect to other
competing environmental values. There may be laws in place or corporate
mantras suggesting sustainable activities, but in reality the former is not enforced and the latter is not practiced (Cerutti et al. 2008).

A large collection of literature critiques the effectiveness of environmental regulation and policy for North America and beyond. In particular,
environmental assessment is international in scope, so researchers have reviewed the characteristics and effectiveness of this ubiquitous process from
a number of national perspectives (e.g., Hanusch and Glasson 2008; Retief
et al. 2008; Samarakoon and Rowan 2008; Zhu and Ru 2008; Wärnbäck
and Hilding-Rydevik 2009). Policy and regulation governing cumulative
effects, the greatest concern for wildlife management and conservation, has
received much attention in Canada (e.g., Dubé and Munkittrick 2001; Noble 2002; Harriman and Noble 2008). Research has shown that at both a
national and a provincial level the assessment and regulation of cumulative
effects, as implemented in the context of environmental assessment and
other regulatory tools, are largely ineffective (Ross 1998; Kennett 1999;
Baxter et al. 2001; Duinker and Greig 2006). Such conclusions are consistent for other jurisdictions, including the United States (Dixon and Montz
1995; Burris and Canter 1997; Spies et al. 2007). Many suggest that cumulative effects must be considered at the regional scale as a component of
proactive long-term land use strategies that are premised on sustainability
(Duinker and Greig 2006). Individual project proposals do not provide adequate scope for assessing future effects that are beyond the interest and
ability of project proponents to identify and quantify. Comprehensive land
use planning and regional environmental assessment are two tools for better incorporating cumulative effects in the decision-making process for the
conservation and management of wildlife populations (Conacher 1994;
Davey et al. 2001; Dubé 2003; Johnson 2011).

Biologists and other natural resource professionals need not play a pas-
sive role in the regulation and mitigation of effects. Indeed, some argue for

a greater role of practitioners and researchers in the critical review and development of policy and law (Meffe and Viederman 1995). Questions of objectivity, the capacity of researchers to accept such roles, and the imposition of research direction through management needs make this a difficult pursuit (Ludwig et al. 2001; Mills and Clark 2001). Less contentious points of integration include research that is designed to explore effective legal thresholds for levels of activity, identifying ecologically meaningful zones of influence or developing processes for mitigation or restoration (e.g., Beckers et al. 2002; Sorensen et al. 2008). Beyond regulatory thresholds, the conservation and management sciences are also important for developing more effective processes and techniques for identifying and integrating impacts in planning and regulatory frameworks (Diaz et al. 2001; Soderman 2006; Mortberg et al. 2007).

Biological Scale of Impacts: A Review

We envision a natural ordering of observed impacts and corresponding methods along a continuum that we have called the biological scale of impact (fig. 3.2). Although these impacts are not discrete, they do correspond to approximate breaks in processes that generally correspond to organismal biology and ecological relationships. This typology is consistent with both hierarchy theory (Allen and Starr 1982) and more progressive definitions of scale, including Dungan et al.'s (2002) scale of phenomena. The role of scale is recognized by others in the context of the evolution of wildlife impact research, but to our knowledge it has not served as a unifying concept across methods, impact types, and taxonomy (Vistnes and Nellemann 2008). Starting at the level of behavioral and physiological systems, we can observe responses that are incrementally more severe and culminate in population-level impacts that can have broader community-level interactions. Although this continuum may be overly simplistic, it does categorize the range of methods used to study the problem and serves as a point of integration to relate the effects of development to the increasing severity of impacts for wildlife.

Assessing Impacts on Individuals Using Changes in Animal Behavior

A behavioral response is probably the first reaction an organism will exhibit to a given modification of its environment, including human-related disturbances (West et al. 2006). From a theoretical perspective, animal behavior is closely linked to ecology and evolution (Krebs and Davies 1993). Although few connections have been made between behavioral ecology and conservation biology in the past (Clemmons and Buchholz 1997), linking behavior, survival, and reproduction may be useful to managers assessing and predicting individual and population-level impacts of human disturbance (Remes 2000; McLoughlin et al. 2005; Haskell and Ballard 2008). Response behaviors result from a combination of proximate factors that may lead to risk aversion or habituation to a given anthropogenic disturbance (Aastrup 2000; Millspaugh et al. 2000; Barten et al. 2001; Borkowski 2001). Studies of animal behavior can be achieved by direct observations (Duchesne et al. 2000), analyses of patterns in data describing distribution (Bergerud and Luttich 2003; Haskell et al. 2006; Preisler et al. 2006), or indirect measures of habitat use, such as browsing by ungulates, (Dahle et al. 2008) and should be performed at several spatial and temporal scales (Romero 2004; Vistnes and Nellemann 2008).

Energy development often has numerous direct and indirect impacts for wildlife that are expressed as changes in specific animal behaviors or activity budgets. Furthermore, direct (e.g., habitat loss) and indirect (e.g., secondary development, change in biological communities) disturbance types may have synergistic effects (Naugle et al. 2011) that are more drastic and long-lasting than the individual parts. Barren-ground caribou and calves avoided road networks near oilfields in Alaska, despite exhibiting adaptable behavior over four decades of disturbance (Haskell and Ballard 2008). Similar avoidance of road networks has been observed in logged landscapes for moose (*Alces alces*; Laurian et al. 2008), elk (*Cervus elaphus*; Fortin et al. 2005), and forest-dwelling caribou (James and Stuart-Smith 2000), probably in relation to wolves (*Canis lupus*), which increase predation efficiency by hunting near roads (Kunkel and Pletscher 2000; Bergman et al. 2006). Linear (e.g., power lines and roads) and punctual disturbances (e.g., oilfield complexes, cabins) have reduced the abundance of wild reindeer (*R. t. tarandus*) in Europe and of caribou in North America (Cameron et al. 1992, 2005; Dyer et al. 2001; Nellemann et al. 2001, 2003; Vistnes and Nellemann 2001; Vistnes et al. 2001, 2004; Dahle et al. 2008). Also, displacement or abandonment of a given area by ungulates often occurs after oil and gas drilling, seismic exploration, and habitat fragmentation (Gillin and Irwin 1985; Kuck et al. 1985; Berger 2004), although only short- to mid-term effects on behavior were sometimes observed. Several studies have documented wildlife avoidance of recreational or tourist structures such as cabins and roads, and such activities as all-terrain vehicle and snowmobile use (Duchesne et al. 2000; Reimers et al. 2003; Taylor and Knight 2003; Preisler et al. 2006; Seip et al. 2007). Such changes in behavior may result in energetic or nutritional costs that constrain population productivity through decreased body weight, conception rate, and age at first reproduction (Reimers et al. 2003). Although the effects of disturbance may be similar between species, the impacts related to wildlife responses are often species-specific; for example, moose often failed to cross pipelines in western Alberta, whereas elk were more successful (Morgantini 1985). Variation in avoidance of a given disturbance also depends on the period of year and the spatial scale considered (Vistnes and Nellemann 2008), highlighting the importance of investigating impacts at several spatiotemporal scales.

Assessing Impacts on Individuals Using Physiology, Energetics, and Nutrition

In conjunction with or after a behavioral interaction with an industrial effect, animals may demonstrate a physiological, energetic, or nutritional response. Depending on the stimulus, physiological responses may take various forms, including increased heart rate and respiration, increased blood flow to skeletal muscle, increased body temperature, and elevated blood sugar level and stress hormone concentration (Gabrielsen and Smith 1995; Fowler 1999; Walker et al. 2005). Over the long term, such physiological responses may be expressed as a decrease in body condition (Amo et al. 2006). For example, fecal corticosterone levels were higher for northern spotted owls (*Strix occidentalis caurina*) that established their home range closer to clearcuts (Wasser et al. 1997). Similarly, the heart rate of mule deer and mountain sheep (*Ovis canadensis mexicana*) drastically increased during simulated low-altitude aircraft noise, but it returned quickly to predisturbance conditions (Weisenberger et al. 1996). From an energetic perspective, some disturbances affect wildlife via increased vigilance, a response that has energetic costs and is akin to the behavior associated with predation risk (Frid and Dill 2002). Also, nutritional and energetic costs may be incurred through altered movement patterns (Johnson et al. 2002) in association with reduced access to foraging or resting habitats. Techniques or measures of changes in body condition, food intake, and stress are numerous. For example, stable isotopes and fatty acids can be used to define spatiotemporal patterns of movement and diet, standardized measurements of body condition are useful for assessing the fitness of individuals within and between populations, and stress hormones and nitrogen metabolites can provide links between animal condition and changes in the environment, including anthropogenic influences (Kofinas et al. 2003; Parker 2003).

The body condition of an animal is an integration of its location-specific energy and nutritional intake and demands, affecting survival, reproduction, and ultimately population dynamics. Indeed, seasonal condition may determine the probability of reproduction, timing of parturition, or survival of neonates (Parker et al. 2009). From an individual perspective, the perception of risk is related to the amplitude of change induced by the disturbance, and the animal's response is proportional to this perceived risk. Consequently, assessing the impacts of disturbance on animal physiology begins with an understanding of the individual's perception of change or risk. After a modification in physiological traits, response to disturbance can be detected at a higher biological scale if the disturbance is severe or prolonged (fig. 3.2).

Population Responses to Cumulative Impacts of Anthropogenic Development

Measures of physiological, energetic, and nutritional responses of individual animals to disturbance are useful, but impacts must ultimately be interpreted at the scale of the population. Indeed, assessing changes in local populations is the key to understanding spatiotemporal population dynamics, evaluating the effectiveness of management strategies for a species, documenting compliance with regulatory requirements, and detecting incipient changes. Abundance may be inferred from direct and indirect estimators, via total counts, sampling a small fraction of a population using standardized indices and estimators, or quantifying indirect evidence of animal presence when individuals are difficult to capture or observe (Gibbs 2000). Considering that methods based on capture can be logistically difficult, time-consuming, and expensive (Hansson 1979; Lancia et al. 1994; Gibbs 2000), researchers and managers need reliable indirect estimators (Slade and Blair 2000), which can be defined as any measurable correlates of density (Caughley 1977). However, using indirect estimators of abundance assumes that the index and actual abundance are related via a positive, linear relationship with slope constant across habitats and over time (Gibbs 2000).

Anthropogenic development can influence population processes in three interrelated ways: changes in population abundance or distribution, modifications in demography, and extirpation. Through research and monitoring, we have gathered much evidence documenting changes in abundance and distribution as a result of disturbance. Wolves with pups tended to abandon recently disturbed areas and move to alternative sites when faced with a repeated disturbance; however, there was no direct relationship with reproductive success (Frame et al. 2007). After the construction of a hydroelectric reservoir in Norway, reindeer gradually reduced use of areas within 4 kilometers of roads and power lines to only 36 percent of the predevelopment densities, and no similar declines were observed in undeveloped control sites (Nellemann et al. 2003). Over the last century, 70 percent of undisturbed reindeer habitat in Norway was lost to piecemeal development.

Disturbance also can induce demographic changes or alter vital rates. Population structure might deviate from natural conditions when impacts are asymmetric across demographic classes including sex, age, and reproductive status. Although measures of abundance are the typical focus of impact studies, changes in reproduction, survival, or other population parameters provide greater mechanistic explanation, are typically easier to observe, and will directly influence population persistence and community dynamics (Gibbs 2000). Declines in reproduction after development or disturbance have been observed in several species, ranging from voles to waterfowl to elephants (Andrews 1990; Madsen 1994, 1998; Forman and Alexander 1998; Lawton et al. 1998; Frid and Dill 2002). The productivity of bald eagles (*Haliaeetus leucocephalus*), for example, declined with increasing proximity of nests to roads; essentially, human presence near nests induced brood abandonment (Trombulak and Frissell 2000). Forman and Alexander (1998) emphasized the importance of increased mortality resulting from vehicle–wildlife collisions. Losses in connectivity between seasonal habitats were noted to affect body condition of female barren-ground caribou with high energy needs during summer (Chan-McLeod et al. 1994), as well as compensatory growth, an essential process to reestablish body weight and ensure subsequent fecundity (Gerhart et al. 1997). When considered as a form of predation risk, human-caused disturbance acts as an evolutionary mechanism linking impacts with animal fitness. A number of studies report the influence of human disturbance on vital rates, including a decline in the survival of cubs after den abandonment by bears (*Ursus* spp.), a reduction in elk calf:cow ratios in areas frequented by tourists, and decreases in reproductive success of mule deer disturbed by all-terrain vehicles (Frid and Dill 2002).

When human-mediated declines in reproduction or increases in mortality reach high levels, range contraction and population extirpation may occur. Forest harvesting has contributed to the extirpation of woodland caribou across the southern fringe of Canadian boreal and subboreal forests (Seip 1992; Schaefer 2003; Courtois et al. 2003; Courtois and Ouellet 2007; Vors et al. 2007; Fortin et al. 2008). Using a geographic comparison of the historic and current distribution of forty-three North American carnivores and ungulates, Laliberte and Ripple (2004) reported that about 40 percent of these species experienced drastic contractions over more than 20 percent of their historic ranges. They concluded that species distribution was more likely to contract and less likely to persist in areas of greater human presence. Similarly, Reid and Miller (1989) reported that human-induced habitat loss is responsible for a large proportion of the extinction events worldwide (fish, 35 percent; birds, 20 percent; mammals, 19 percent; reptiles, 5 percent) and currently threatens many other species.

Community Responses to Cumulative Impacts of Anthropogenic Development

Anthropogenic disturbance can drastically alter community structure by influencing the underlying processes of interspecific relationships. The mechanisms of such interactions are community-specific but may result in the modification of rates or linkages between species involved in herbivory, predation, parasitism, competition, mutualism, or commensalism. Disturbance of benthic ocean communities by trawling may be modest at the population scale (e.g., decreases in abundance and species richness), but the cumulative impacts can induce profound changes at the community level by compromising the integrity of marine food webs (Hinz et al. 2009). Studying the impacts of mining and recreation on relationships between wolves and barren-ground caribou in the Northwest Territories, Frame et al. (2008) observed that wolf reproductive success was lower during years when migration routes of caribou were farther from wolf dens, thus forcing wolves to travel farther to access their main prey. Longer periods of time spent traveling between caribou herds and dens decreased net energy intake for pups and lowered their probability of survival. In response to road development and human-induced disturbance in the Yellowstone Ecosystem, moose synchronized calving events within a 9-day period and selected habitats close to paved roads in order to shift away from traffic-averse brown bears (Berger 2007). This behavioral response reduced predation risk for neonates but concurrently reduced the hunting success of bears, with potential implications for fitness.

Wildlife–habitat relationships are premised on the optimization of individual fitness, where animals select habitats that maximize opportunities to reproduce and increase their probability of survival (Remes 2000; McLoughlin et al. 2005). For a given species, the choice of habitat probably is based on the recognition of cues, such as forage quality and forest structure, that relate to increased fitness (Ratti and Reese 1988; Purcell and Verner 1998). However, after sudden changes in the environment, such as anthropogenic disturbance, these same cues could lead to bad habitat choices or behavioral asynchrony (Best 1986; Boal and Mannan 1999). Population declines may then occur within these ecological traps: modified habitats with characteristics that appear to offer an advantage but in reality provide suboptimal conditions for reproduction or survival (Schlaepfer et al. 2002; Battin 2004). Examples of ecological traps from the avian literature include high nesting mortality in urban areas due to nest parasites (Boal and Mannan 1999) and higher nest predation near forest–power line edges (Chasko and Gates 1982) or forest–clearcut edges (Flaspohler et al. 2001).

The decline and extirpation of woodland caribou across Canada is an excellent example of the implications of anthropogenic developments for community-level processes (Seip 1992; Wittmer et al. 2005; Courtois and Ouellet 2007). Across boreal and subboreal forests, widespread logging results in mature forest being replaced by early successional species that are suitable winter habitat for moose (Osko et al. 2004; Dussault et al. 2005). Wolves prey on both moose and caribou, but their populations are regulated by moose density. As wolves increase in distribution and abundance relative to expanding moose populations, caribou decline as a secondary and nonregulating prey item, a phenomenon called apparent competition (Bergerud and Elliot 1986; Seip 1992; Wittmer et al. 2005). Regenerating forests are also rich in food resources for black bears (e.g., berries; Schwartz and Franzmann 1989; Brodeur et al. 2008), which may influence caribou recruitment through predation on calves (Adams et al. 1995; Linnell et al. 1995). As logging creates younger forests across a larger proportion of managed landscapes, caribou are forced to frequent remnant old stands that are adjacent to high-quality habitats for moose, wolves, and bears (Hins et al. 2009). This increased proximity between risk-prone and suitable habitats ultimately creates an ecological trap, thus compromising the evolved antipredator strategies of caribou to "space away" from predators.

At an even higher level of ecological complexity, anthropogenic disturbances can have important cascading effects on wildlife community structure and dynamics. Working in Banff National Park, Alberta, Canada,

Hebblewhite et al. (2005b) measured the responses of plant and animal populations to a range of densities of wolves, the dominant predator in the system. When occurring at low density, as a result of human activities, wolves were unable to regulate elk populations and consequently to limit herbivory on willow (*Salix* spp.) and aspen (*Populus tremuloides*). Through high grazing pressure, elk limited the recruitment of willow and aspen, effectively outcompeting beaver and negatively affecting the diversity and abundance of riparian songbirds.

Assessing Impacts: Choosing Methods and Scales

Choice of methods is extremely important when studying the influence of human development on wildlife. Experimental design is important for any scientific investigation, but in the case of wildlife impact studies, a conclusion premised on a false positive or false negative could compromise the persistence of a wildlife population, evaluation of mitigation strategies, or economic development (Gibbs et al. 1998).

Simply monitoring indices or direct measures of impacts (e.g., abundance) does not guarantee results that will warrant a regulatory, management, or conservation response (see Gibbs 2000 for a review of common pitfalls in monitoring programs). Designs can lack sufficient statistical power to detect population trends (chap. 5), and monitoring programs often fail to account for natural variability that may mask the disturbance response of interest (Gibbs et al. 1998; Gibbs 2000). For experimental design, power analyses can be useful for identifying the appropriate level of replication or duration of a study (chap. 5). Without multiyear investigations or preliminary pilot data, however, biologists often have no idea of the variance associated with the assumed response or the underlying ecological processes. Alternatively, one can rely on measures of effect size to guide experimental design (Yoccoz 1991; Cohen 1994; Kirk 1996). Large responses (i.e., effect sizes) can be detected reliably with few sites (or plots) and infrequent monitoring; however, comparatively slight changes in animal responses will necessitate a larger number of sites and a longer duration of monitoring (Gibbs et al. 1998). Even with large effect sizes, natural variability in population-level processes could mask underlying trends resulting from disturbance responses. Thus, depending on the life history strategy of the species of interest, the magnitude of impact, and possible lag effects, ecologically significant trends may become apparent only at time spans greater than the duration of typical research projects (i.e., more than 2 years).

Choice of Methods

The choice of methods, biological scales, and even model species for impact studies is not guided by statistical criteria alone. The assumed type and magnitude of impact; past research on the population, species, or development type; and even regulatory framework can influence the focus and design of a study. Furthermore, we may expect a shifting bias in the types of questions and species that are investigated. Research on the impacts of development for caribou and reindeer has evolved from studies of behavioral responses of individuals to population-level processes (Vistnes and Nellemann 2008). Similarly, researchers and land managers now have a greater appreciation for the cumulative impacts of development (e.g., Schneider et al. 2003; Johnson et al. 2005; fig. 3.1). As an example of taxonomic bias, one need only look at studies focused on the impacts of wind energy. Here, birds and bats (as flying species are more prone to collision) or marine species (for offshore wind farms) are the model species (e.g., Gill 2005; Köller et al. 2006; Madsen et al. 2006). Observed impacts are both direct, such as fatalities caused by collisions with turbines, and indirect, such as the disruption of foraging behavior, breeding activities, and migratory patterns due to landscape alteration around wind farms (Erickson et al. 2001; Kunz et al. 2007a, 2007b; Arnett et al. 2008). Unfortunately, very few studies have looked at the impacts of wind energy for terrestrial mammals, amphibians, reptiles, or insects (but see de Lucas et al. 2005; Rabin et al. 2006). Consequently, most of the methods tested for this particular energy source are focused exclusively on avifauna (e.g., netting, thermal infrared imaging, radar monitoring, acoustic recordings, and fatality counts).

We have identified three general sets of methods that have demonstrated utility for wildlife impact studies. First, impacts can be studied experimentally, where one or several disturbances can be applied to treatment and control group. This strategy is powerful when conducted at a small spatial scale, allowing researchers to control confounding factors (e.g., habitat type, food resources, abundance of competitors) that may influence a response to disturbance. At large spatial scales, however, such designs are difficult to implement because of biological, logistical, and financial constraints that result in small sample sizes and heterogeneity within treatment units (Roedenbeck et al. 2007). In a review on bird habitat use in logged landscapes, Sallabanks et al. (2000) reported that only 27 percent ($N = 25$) of reviewed studies had more than four replicates. In a literature review of the impacts of energy development on ungulates, none of the 120 publications reviewed were replicated at the level of impact type, and all studies had only one replicate (chap. 5). This author presented many examples of experiments that were constrained by insufficient statistical power, thus increasing Type II error.

Natural variability in treatment units can obscure impacts and negate the inherent strengths of experimental design (St-Laurent et al. 2007, 2008). Simply put, underlying uncontrolled habitat structure and composition may have a larger influence on species behavior, physiology, distribution, or abundance than treatment type. This source of confounding variation may lead to a false conclusion that disturbances had no impact (i.e., Type II error).

When controlled application of disturbance effects and replication is impractical, an alternative and simpler approach is to monitor free-ranging animal populations exposed to naturally varying levels of disturbance. Observational assessments are based on uncontrolled systems, where confounding variation occurs and should be taken into account often through statistical method, not study design. Here, levels of disturbance are not fixed by researchers but are constrained by the existing availability of anthropogenic footprints. Although potentially less expensive and challenging from a logistics perspective, such a study design can suffer from two important pitfalls. First, many variables must be considered in order to isolate the impact of disturbance because their respective effects could be habitat-mediated (Gibbs et al. 1998; Gibbs 2000). Second, the level of impact would be restricted to the observational range of disturbance intensity. Inference and prediction would then be restricted, especially in environments where the disturbance footprint is still low. This limitation was demonstrated by Johnson et al. (2005) when they attempted to measure the impacts of mineral development on the distribution of arctic mammals. Here, a large undisturbed landscape with few point sources of disturbance naturally limited the magnitude of impact and constrained any potential conclusions about possible future outcomes of increased development intensity. Examples of observational impact studies are numerous, from using radio or Global Positioning System (GPS) collars to monitor the distribution of free-ranging animals inhabiting modified landscapes, to observations of animal behavior or physiological measures under existing levels of human development (e.g., St-Laurent et al. 2007; Thiel et al. 2008; Hins et al. 2009).

Finally, simulation models have become an increasingly important tool for understanding wildlife–impact relationships. In contrast to other research methods, these models have the capacity to forecast the magnitude of impacts over broader biological, spatial, and temporal frameworks. When coupled with hypothetical development scenarios (Peterson et al.

2003; Duinker and Greig 2007) and stakeholder participation, simulation models are an effective device for exploring and reaching consensus on the sustainability of human activities over broad regional areas or planning horizons. These models serve an equally important role as catalysts for integrating knowledge and generating pertinent hypotheses. The process of building the model forces the research team not only to explore and formalize linkages between processes they understand, but also to document processes and mechanisms where knowledge is lacking (Starfield 1997). Wildlife researchers and natural resource professionals have used simulation models to address impacts at a number of biological scales. Linking behavior, energetics, and ultimately the reproductive capacity of the population, Russell et al. (2005) drew on a large body of independent research to generate a mechanistic model that predicts the daily weight of female barren-ground caribou and dependent calf in the context of climate change and the development of oil reserves. Often, however, the necessary data describing the links between behavior, physiology, and population dynamics are unavailable. Less mechanistic approaches have related broad measures of distribution, habitat needs, or vital rates to human disturbance. Population viability analysis (PVA) has served as the general modeling paradigm for many of these efforts. Essentially a stochastic population model, PVA is inherently flexible, permitting a nearly infinite range of life history strategies and development or disturbance scenarios (Boyce 1992). For example, PVA was used to quantify the relationship between the probability of persistence of a population of Eurasian corvid (*Pyrrhocorax pyrrhocorax*) and declines in juvenile survival related to visitation rates of tourists (Kerbiriou et al. 2009). A population model for woodland caribou related landscape disturbance from oil and gas activities to a general measure of forage availability, intake, and predation (Weclaw and Hudson 2004).

Simulation models have shown great promise in addressing cumulative effects. For example, A Landscape Cumulative Effects Simulator (ALCES) has been used in numerous jurisdictions across western Canada as a stakeholder engagement and facilitation tool that tracks important indicators of sustainability, including wildlife habitat, and predicts the outcomes of development scenarios (Schneider et al. 2003). Examples of project-specific analyses of cumulative effects include the work of Johnson et al. (2005) in the Northwest and Nunavut territories of Canada. Using existing presence data for four species of arctic mammal, they quantified the additive distribution and habitat implications of future increases in mineral development. Focusing on population demographics, Nielsen et al. (2006) used models of occurrence and mortality risk to rank five habitat states related to sink–source dynamics for grizzly bears in western Alberta, Canada. Ya-masaki et al. (2008) adopted a more holistic approach, developing a land-scape disturbance model that predicted the impacts of forestry, oil and gas, climate change, and fire for key indicators of forest productivity and bio-diversity. Relative to other efforts, their model is noteworthy because it captured the interactions between several disturbance factors, an important element of cumulative effects. Numerous other recent studies illustrate species-specific or broader ecosystem-based cumulative effect assessments for wildlife populations (e.g., Gustafson et al. 2007; Harvey and Railsback 2007; Swenson and Ambrose 2007; Shifley et al. 2008). Although these are useful tools, results may be limited by the range of variation of the dif-ferent biological rates included in the model and by the model elasticity (Mills et al. 1999).

In addition to the choice of research technique and model species, re-searchers must determine the most appropriate biological scale of study. We argue that these scales are permeable, not discrete, and that the range of im-pacts across scales should be an explicit consideration during reporting if not during study design. For example, a decrease in foraging opportunities within an animal's home range will lead to an increase in movement rate and a decrease in nutritional plane, thus limiting the energy available to in-vest in reproduction. The implications of the impacts for the individual fol-low logically for the population and community if a significant proportion of the females do not achieve their reproductive potential and the species is integral to other trophic interactions. A common pitfall in studying impacts of a given disturbance is that experiments are conducted on a small scale and results may be inaccurate for extrapolation to the larger temporal, spa-tial, or biological scale (Vistnes and Nellemann 2008), especially when ani-mals have been confined to experimental structures in which some impor-tant ecological factors can be modified or ignored.

Conclusion

We outlined a typology that is simple but represents a set of concepts that allow researchers and managers to compare the results of individual studies across taxonomy, time, and space. This is an important first step in improv-ing the generality of wildlife impact science, management, and conserva-tion. When the biological scale of impacts is combined with an understand-ing of the range of effects represented by the three axes in figure 3.2, resource professionals have a tool to relate effects to the observed or predicted impacts. Thus, the typology can serve as a starting point for comparing results of previously divergent studies or predict the potential magnitude of impacts before resource development proceeds, based on the range of expected effects. If nothing else, the framework can serve as a focal point for developing a common language based on the science of wildlife impact research. Such unification and dialogue is greatly needed, as illustrated by the 40 years of impact research for barren-ground caribou (*R. t. granti*) calving near the Prudhoe Bay oil facility. The resulting acrimonious dialogue between competing camps clearly demonstrates that our interpretation and communication of the relevant science must be improved (Cronin et al. 2000; Cameron et al. 2005; Haskell et al. 2006; Joly et al. 2006; Noel et al. 2004, 2006).

We provided some perspective on general methods that may be applicable to wildlife impact studies. Guidance on general sampling and research design is well represented in the literature (e.g., Krebs 1999; Gibbs 2000). However, we argue that challenges for researchers conducting impact studies are unique, especially where the focus is on higher-level processes across broad spatiotemporal scales. As a starting point, researchers should assume a holistic perspective that implicitly or explicitly considers the range of biological or ecological scales. From the perspective of individual research projects, the investigator is then challenged with the choice of biological scale for study. One can provide the standard blanket proviso that the study metric designed to quantify the level of impact should depend on the development activity, species of interest, or availability of methods. Instead, we suggest a more thoughtful approach. The scale of impact and supporting measures or monitoring strategies should be chosen relative to the expected level of effect. If we expect a small effect based on observation and application of figure 3.2, then it makes little sense to attempt to measure impacts relative to higher-level processes, such as population or community dynamics. Likewise, where effects are cumulative, have a large spatiotemporal footprint, and are unconstrained by legislation or management, we may fail to record important higher-level responses if we focus our studies on trivial impacts, such as the behavioral or physiological responses of individuals. All is not lost when the appropriate biological scale of impact is undermatched, but where further development decisions must be made quickly, funding for research and monitoring is limited, or a population is imperiled, the implications of incorrect study design may be substantial. In the absence of guidance, or in the face of uncertainty in the anticipated level of effect, we should assume that impacts are likely to be more significant and difficult to reverse as we move up the biological, spatiotemporal, and management gradients outlined in figure 3.2. Although understanding the impacts of human developments at all biological scales may be an interesting, integrative, and preventive assessment strategy, we agree with Vistnes and Nellemann (2008) that researchers should pay more attention to population-level and community-level impacts where appropriate.

Some may argue that we need to improve the integration of research findings with the management and mitigation of effects, or improve study design so that research questions directly address management questions (chap. 5). Although we agree that this is a worthwhile objective, we think that the next step for the evolution of impact research is synthesis. Our review suggests a large and increasing body of studies and knowledge on the level and types of impacts for a wide range of species (fig. 3.1). Given current rates of exploration and development of energy sources, we do not anticipate a decrease in such efforts. Unfortunately, comparisons of findings across taxonomic groupings or types of development effects were lacking from this impressive and well-funded set of studies. We suggest a much greater commonality between these supposedly divergent studies and impacts than is represented in formal dialogue. Coordination and communication are essential if we are to develop a body of research techniques and understanding that mitigates or reduces the cumulative, species-specific, and community-level impacts of resource development. Guidance on such efforts cannot be expected to come from legislators or developers but rather must come from the scientists, managers, and conservationists who understand the implications of impacts for the distribution and abundance of wildlife.

Chapter 4

Sage-Grouse and Cumulative Impacts of Energy Development

DAVID E. NAUGLE, KEVIN E. DOHERTY,
BRETT L. WALKER, HOLLY E. COPELAND,
MATTHEW J. HOLLORAN, AND JASON D. TACK

World demand for energy increased by more than 50 percent in the last half-century, and a similar increase is projected between now and 2030 (National Petroleum Council 2007). Fossil fuels will remain the largest source of energy worldwide, with oil, natural gas, and coal accounting for more than 80 percent of world demand (chap. 1). Projected growth in U.S. energy demand is 0.5–1.3 percent annually (National Petroleum Council 2007), and development of domestic reserves will expand through the first half of the twenty-first century. Western states and provinces will continue to play a major role in providing additional domestic energy resources to the United States and Canada, which is expected to place unprecedented pressure on the conservation of wildlife populations throughout the West.

The sagebrush ecosystem is representative of the struggle to maintain biodiversity in a landscape that bears the burden of our ever-increasing demand for natural resources. One species affected by domestic energy production is the greater sage-grouse (*Centrocercus urophasianus*; hereafter "sage-grouse"), a game bird endemic to western semiarid sagebrush (*Artemisia* spp.) landscapes in western North America (Schroeder et al. 1999). Previously widespread, the sage-grouse has been extirpated from approximately half of its historic range (Schroeder et al. 2004), and populations have declined by 1.8–11.6 percent annually over the past four decades in about half of the populations studied (Garton et al. 2011). Energy development has emerged as a major issue in conservation because areas currently under development contain some of the highest densities of sage-grouse (Connelly et al. 2004) and other sagebrush obligate species (Knick et al. 2003) in western North America.

The sage-grouse is considered a landscape species in that it needs large, intact sagebrush habitats to maintain robust populations (Connelly et al. 2011). As a result, the size of sage-grouse breeding populations is often used as an indicator of the overall health of the sagebrush ecosystem (Hanser and Knick 2011). There are few early studies evaluating the impacts of energy development on sage-grouse populations (see Naugle et al. 2011), but research has increased rapidly in concert with the pace and extent of development. The goal of this chapter is to provide a scientific understanding of impacts of energy development on sage-grouse and to recommend biologically based solutions. Objectives are to synthesize current data regarding the biological response of sage-grouse to energy development, identify ecological and behavioral mechanisms causing population-level impacts, evaluate empirically the extent to which current and anticipated development affects populations, and outline a strategy for landscape conservation analogous in scale to that of ongoing and anticipated impacts of development.

Biological Response of Sage-Grouse to Energy Development

We searched the literature for studies that investigated relationships between sage-grouse and energy development using methods described in Naugle et al. (2011). We included theses and dissertations but excluded from review documents that included only cautionary statements about potential impacts or anecdotal data. Fourteen studies reported negative impacts of energy development on sage-grouse (table 4.1).

None of the fourteen studies synthesized here reported a positive influence of development on populations or habitats.

Breeding populations were severely affected at well densities commonly permitted (eight pads per 2.6 square kilometers) in conventional oil and gas fields in Montana and Wyoming (Holloran 2005; Walker et al. 2007a). Magnitude of losses may vary from one field to another, but findings to date suggest that impacts are universally negative and typically severe. Surface occupancy of oil or gas wells adjacent to leks was negatively associated with male lek attendance in five of seven study areas across Wyoming (Harju et al. 2010). Leks with at least one oil or gas well within a 0.4-kilometer radius had 35–91 percent fewer attending males than leks

TABLE 4.1. Research studies on effects of oil and gas development on greater sage-grouse.

Citation and Study Location	Research Outlet	Pretreatment Design	Pretreatment/ Control	Years	Sample Size
Lyon and Anderson (2003), Pinedale Mesa, SW Wyoming	Scientific journal	Observational	N/Y	2	48 females from 6 leks
Holloran (2005), Pinedale Anticline Project Area and Jonah II gas field, SW Wyoming	Ph.D. dissertation	Correlative and observational	N/Y	7	Counts of 21 leks, 209 females from 14 leks, 162 nests
Kaiser (2006) Pinedale Anticline Project Area and Jonah II gas field, SW Wyoming	M.S. thesis	Correlative and observational	N/Y	1	18 leks, 83 females (23 yearlings), plus 20 yearling males
Holloran et al. (2007), Pinedale Anticline Project Area and Jonah II gas field, SW Wyoming	U.S. Geological Survey	Correlative and observational	N/Y	2	86 yearlings (52 females), 23 yearlings (17 females) with known maternity
Aldridge and Boyce (2007), SE Alberta, Canada	Scientific journal	Correlative	N/N	4	113 nests, 669 locations on 35 broods, 41 chicks from 22 broods
Walker et al. (2007a), Powder River Basin, NE Wyoming and SE Montana	Scientific journal	Correlative	Y/Y	8	97–154 lek complexes/year for trends, 276 lek complexes in status analysis

TABLE 4.1. Continued

Citation and Study Location	Research Outlet	Pretreatment Design	Pretreatment/ Control	Years	Sample Size
Doherty (2008), Powder River Basin, NE Wyoming and SE Montana	Ph.D. dissertation	Correlative and observational	N/Y	10	1,190 active leks and 154 inactive leks
Doherty et al. (2008), Powder River Basin, NE Wyoming and SE Montana	Scientific journal	Correlative	N/N	4	435 locations to build model, 74 new locations from different years to test it
Tack (2009), Montana, SE Alberta, SW Saskatchewan, SW North Dakota, and NW South Dakota	M.S. thesis	Correlative and observational	N/N	10	802 active and 297 inactive lek complexes
Holloran et al. (2010), Pinedale Anticline Project Area and Jonah II gas field, SW Wyoming	Scientific journal	Correlative and observational	N/Y	6	135 yearling females and 34 yearling males, from 17 leks, and 62 yearling sage-grouse nests
Harju et al. (2010), 7 sites in Wyoming	Scientific journal	Correlative and observational	N/Y	12	704 leks; 4,861 lek-year data points
Doherty et al. (2010a)	Scientific journal	Correlative and observational	N/Y	11	1,344 leks
Carpenter et al. (2010)	Scientific journal	Observational	N/N	4	296 locations for 23 females
Johnson et al. (2011), range-wide	Scientific journal	Correlative	N/N	11	3,679 leks with at least 4 counts from 1997 to 2007

with no well within this radius (Harju et al. 2010). Declining lek attendance was also associated with a higher landscape-level density of well pads; lek attendance at well pad densities of 1.54 well pads per square kilometer was 13–74 percent lower than attendance at nonimpacted leks (leks with zero well pads within 8.5 kilometers) (Harju et al. 2010). Lek attendance at a well pad density of 3.1 well pads per square kilometer was 77–79 percent lower than attendance at leks with no well pad within 8.5 kilometers. Similar analyses throughout Wyoming showed no difference in lek persistence at one well pad per 2.6 square kilometers (Doherty 2008; Doherty et al. 2010a), but declines in males at large leks (more than twenty-five males) were apparent at less than one well pad per 2.6 square kilometers in eastern Montana (fig. 9 in Tack 2009).

Negative impacts are known for four different sage-grouse populations in three different types of development, including shallow coal bed natural gas in the Powder River Basin of northeast Wyoming and extreme southeast Montana (Walker et al. 2007a; Doherty et al. 2008), deep gas in the Pinedale Anticline Project Area in southwest Wyoming (Lyon and Anderson 2003; Holloran 2005; Holloran et al. 2010), oil extraction in the Manyberries Oil Field in southeast Alberta (Aldridge and Boyce 2007), and oil and gas development in the Cedar Creek Anticline near the state boundaries between southeast Montana, western North Dakota, and northwestern South Dakota (Tack 2009). Population trends in the Powder River Basin indicated that from 2001 to 2005, lek count indices inside gas fields declined by 82 percent, whereas indices outside development declined by 12 percent (fig. 4.1).

Of leks active in 1997, only 38 percent inside gas fields remained active as of 2004–2005, compared with 84 percent outside development (Walker et al. 2007a). Male lek attendance in the Pinedale Anticline decreased with distance to the nearest active drilling rig (fig. 4.2), producing gas well, and main haul road, and declines were most severe (40–100 percent) at breeding sites within 5 kilometers of an active drilling rig or within 3 kilometers of a producing gas well or main haul road (Holloran 2005).

In an endangered population in Alberta, Canada, where low chick survival (12 percent to 56 days) limits population growth, risk of chick mortality in the Manyberries Oil Field was 1.5 times higher for each additional well site visible within 1 kilometer of a brood location (Aldridge and Boyce 2007). At the Cedar Creek Anticline in southeast Montana, male abundance at leks decreased by 52 percent at sixteen leks with more than one well pad per 2.6 square kilometers within 3.2 kilometers, and no males

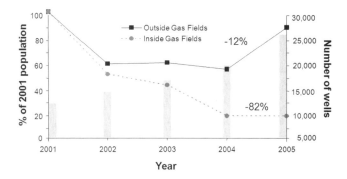

FIGURE 4.1. Population indices based on male lek attendance for sage-grouse in the Powder River Basin, Montana and Wyoming, 2001–2005 for leks categorized as inside or outside coal bed natural gas fields on a year-by-year basis (as modified from Walker et al. 2007a). Leks in gas fields had at least 40% energy development within 3.2 kilometers or more than 25% development within 3.2 kilometers and at least 1 well within 350 meters of the lek center. Number of producing gas wells in the basin shows the overall increase in development coincident with declines in sage-grouse population indices.

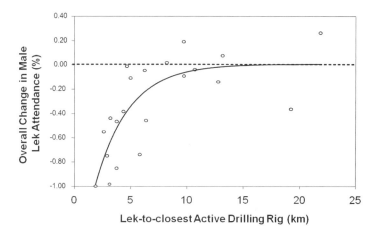

FIGURE 4.2. Relationship between number of sage-grouse males attending leks and average distance from leks to closest active gas drilling rig, Pinedale Anticline Project Area, southwest Wyoming, 1998–2004. Each point along the regression line represents one lek ($N = 21$; modified from Holloran 2005).

were counted in 2009 at four of the sixteen impacted leks that had multiple displaying males in 2008 (Tack 2009).

Studies also have quantified the distance from leks at which impacts of development become negligible, and have assessed the efficacy of the stipulation by the U.S. Bureau of Land Management (BLM) of no surface disturbance within 0.4 kilometers of a lek. Impacts to leks from energy development were most severe near the lek, remained discernible out to distances of more than 6 kilometers (Holloran 2005; Walker et al. 2007a; Tack 2009; Johnson et al. 2011), and often resulted in extirpation of leks within gas fields (Holloran 2005; Walker et al. 2007a). Negative effects of well surface occupancy were apparent out to 4.8 kilometers, the largest radius investigated, in two of seven study areas in Wyoming (Harju et al. 2010). Curvilinear relationships showed that lek counts decreased with distance to the nearest active drilling rig, producing well, or main haul road, and that development within 4.7–6.2 kilometers of leks influenced counts of displaying males (fig. 4.2; Holloran 2005). All well-supported models in Walker et al. (2007a) indicated a strong negative effect of energy development, estimated as proportion of development within either 0.8 kilometer or 3.2 kilometers, on lek persistence. A model with development at 6.4 kilometers had less support (5–7 ΔAICc units lower), but the regression coefficient (β = –5.11, SE = 2.04) indicated that negative impacts of development within 6.4 kilometers were still apparent. Walker et al. (2007a) used the resulting model to demonstrate that the 0.4-kilometer lease stipulation used by the BLM was insufficient to conserve breeding sage-grouse populations in fully developed gas fields. A 0.4-kilometer buffer leaves 98 percent of the landscape within 3.2 kilometers of a lek open to full-scale development. Full-field development of 98 percent of the landscape within 3.2 kilometers of leks in a typical landscape in the Powder River Basin reduced the average probability of lek persistence from 87 percent to 5 percent (Walker et al. 2007a). Two recent studies found negative impacts apparent out to 12.3 kilometers on large lek occurrence (more than twenty-five males; Tack 2009) and out to 18 kilometers on lek trends (Johnson et al. 2011), the largest scales that have yet been evaluated.

Negative responses of sage-grouse to energy development were consistent among studies, regardless of whether they examined lek dynamics or demographic rates of specific cohorts within populations. Sage-grouse populations decline when birds avoid infrastructure in one or more seasons (Doherty et al. 2008; Carpenter et al. 2010) and when cumulative impacts of development negatively affect reproduction or survival (Aldridge and

Boyce 2007) or both demographic rates (Lyon and Anderson 2003; Holloran 2005; Holloran et al. 2010). Avoidance of energy development at the scale of entire oil and gas fields should not be considered a simple shift in habitat use but rather a reduction in the distribution of sage-grouse (Walker et al. 2007a). Avoidance is likely to result in true population declines if density dependence, competition, or displacement of birds into poorer-quality adjacent habitats lowers survival or reproduction (Holloran and Anderson 2005; Aldridge and Boyce 2007; Holloran et al. 2010). High site fidelity in sage-grouse also suggests that unfamiliarity with new habitats may also reduce survival, as in other grouse species (Yoder et al. 2004). Sage-grouse in the Powder River Basin were 1.3 times more likely to occupy winter habitats that had not been developed for energy (twelve wells per 4 square kilometers), and avoidance of developed areas was most pronounced when it occurred in high-quality winter habitat with abundant sagebrush (Doherty et al. 2008). In a similar study in Alberta, avoidance of otherwise suitable wintering habitats within a 1.9-kilometer radius of energy development resulted in substantial loss of functional habitat surrounding wells (Carpenter et al. 2010). Authors recommend at least a 1,900-meter setback distance for future energy developments from all winter habitats identified as Critical Habitat under the federal Species at Risk Act for this endangered species in Canada.

Long-term studies in the Pinedale Anticline Project Area in southwest Wyoming present the most complete picture of cumulative impacts and provide a mechanistic explanation for declines in populations. Early in development, nest sites were farther from disturbed than undisturbed leks, the rate of nest initiation from disturbed leks was 24 percent lower than for birds breeding on undisturbed leks, and 26 percent fewer females from disturbed leks initiated nests in consecutive years (Lyon and Anderson 2003). As development progressed, adult females remained in traditional nesting areas regardless of increasing levels of development, but yearlings that had not yet imprinted on habitats inside the gas field avoided development by nesting farther from roads (Holloran 2005). The most recent study confirmed that yearling females avoided infrastructure when selecting nest sites, and yearling males avoided leks inside of development and were displaced to the periphery of the gas field (Holloran et al. 2010). Recruitment of males to leks also declined as distance within the external limit of development increased, indicating a high likelihood of lek loss near the center of developed oil and gas fields (Kaiser 2006).

The most important finding from studies in Pinedale was that sage-grouse declines are explained in part by lower annual survival of female sage-grouse and that the impact on survival resulted in a population-level decline (Holloran 2005). However, we still lack a clear picture of long-term effects of behavioral avoidance coupled with decreased survival. The population decline observed in sage-grouse is similar to that observed in Kansas for the lesser prairie chicken (*Tympanuchus pallidicinctus*; Hagen 2003), a federally threatened species that also avoided otherwise suitable sand sagebrush (*Artemisia filifolia*) habitats proximal to oil and gas development (Pitman et al. 2005; Johnson et al. 2006). High site fidelity but low survival of adult sage-grouse, combined with lek avoidance by younger birds (Holloran et al. 2010), resulted in a time lag of 3–4 years between the onset of development activities and lek loss (Holloran 2005). The time lag observed by Holloran (2005) in the Anticline matched that for leks that became inactive 3–4 years after natural gas development in the Powder River Basin (Walker et al. 2007a). Analysis of seven oil and gas fields across Wyoming showed time lags of 2–10 years between activities associated with energy development and its measurable effects on sage-grouse populations (Harju et al. 2010). This knowledge of time lags suggests that ongoing development in the Cedar Creek Anticline will result in additional impacts on fringe populations in eastern Montana and western North and South Dakota (Tack 2009).

Mechanisms that lead to avoidance and decreased fitness have not been empirically tested but are suggested from observational studies. For example, abandonment may increase if leks are repeatedly disturbed by raptors perching on power lines near leks (Ellis 1984), by vehicle traffic on nearby roads (Lyon and Anderson 2003), or by noise and human activity associated with energy development during the breeding season (Holloran 2005; Kaiser 2006). Collisions with nearby power lines and vehicles and increased predation by raptors may also increase the mortality rate of birds at leks (Connelley et al. 2000a). Alternatively, roads and power lines may indirectly affect lek persistence by altering the productivity of local populations or survival at other times of the year. For example, sage-grouse mortality associated with power lines and roads occurs year-round (Patterson 1952; Beck et al. 2006; Aldridge and Boyce 2007), and ponds created by coal bed natural gas development may increase the risk of West Nile virus mortality in late summer (Walker et al. 2004; Zou et al. 2006; Walker et al. 2007b). Loss and degradation of sagebrush habitat can also reduce the carrying capacity of local breeding populations (Swenson et al. 1987; Braun 1998; Connelly et al. 2000a, 2000b; Crawford et al. 2004). Alternatively, birds may simply avoid otherwise suitable habitat as the density of roads, power lines, or energy development increases (Lyon and Anderson 2003; Holloran 2005; Kaiser 2006; Doherty et al. 2008; Carpenter et al. 2010).

A Shifting Paradigm: Business as Usual versus Landscape Conservation

The unequivocal answer to energy development in the West is not "no" but rather "where" to reduce environmental impacts while still extracting resources to meet increasing domestic demand for energy. The U.S. government has already leased more than 7 million hectares of the federal mineral estate, and the number of producing wells tripled from 11,000 in the 1980s to more than 33,000 in 2007 (Naugle et al. 2011). Managers struggling to keep pace with development have implemented reactive measures in hopes of mitigating disturbance around leks. Protective measures, such as not allowing energy infrastructure within varied distances around leks and timing restrictions on drilling, have failed to maintain populations, and it has become apparent that sage-grouse conservation and energy development are incompatible in the same landscapes.

Budgetary constraints to study and maintain wildlife populations mean that conservation triage is unavoidable, defined here as the prioritization of limited resources to maximize biological return on investment (Bottrill et al. 2008), to meet the high social and economic costs of maintaining sage-grouse populations. The focus for conservation should be to prioritize and conserve remaining intact landscapes rather than trying to maintain small declining populations at the cost of further loss in the best remaining areas (chap. 12). The challenge will be to implement conservation on a scale that matches energy development to offset the spatial extent of anticipated impacts. Scientists need to work with managers to develop proactive decision support tools that identify priority landscapes that will maintain large populations, develop management prescriptions that increase populations in priority landscapes and offset losses in developed landscapes, and identify ecological corridors among priority populations to maintain connectivity. Despite ongoing development, no comprehensive rangewide plan is in place to conserve large and functioning landscapes to maintain sage-grouse populations.

Conservation Planning Using Core Areas to Reduce Impacts

Analytical frameworks are available to evaluate options for reducing impacts on sage-grouse populations at highest risk of oil, gas, and wind power development. For example, Doherty et al. (2011) used lek count data ($N =$ 2,336 leks) to delineate high-abundance population centers, or core areas, that contained 25, 50, 75, and 100 percent of the known breeding populations in Wyoming, Montana, Colorado, Utah, and North and South Dakota (fig. 4.3).

Core areas can be overlaid spatially with authorized oil and gas leases and the potential for commercial development of wind energy (fig. 4.4), and the resulting output can then be used to identify the least at-risk core populations to energy development to prioritize for immediate conservation.

Areas that share high energy potential and high sage-grouse density will need policy reform to reduce threats, whereas areas with high energy

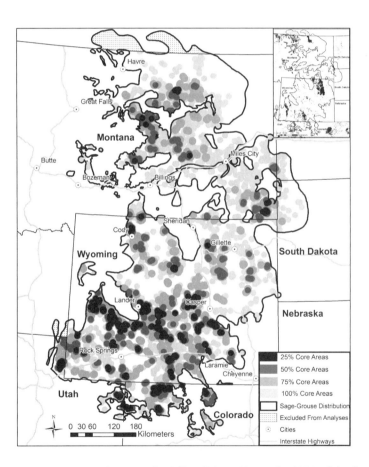

FIGURE 4.3. Core areas that contain 25%, 50%, 75%, and 100% of the known breeding population of greater sage-grouse in their eastern range (modified from Doherty et al. 2011). Distribution boundaries are the combined areas of sage-grouse management zones I and II (Connelly et al. 2004). Inset depicts locations of producing oil and gas wells (black) as of September 2007.

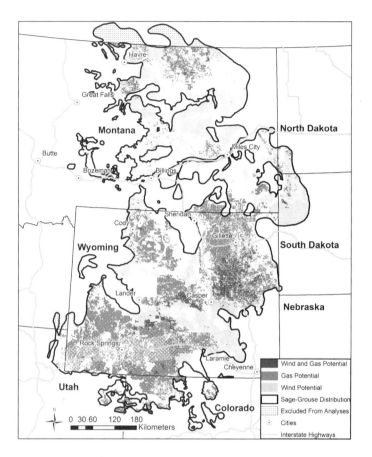

FIGURE 4.4. Potential for oil, gas, and wind power development in the eastern range of greater sage-grouse (management zones I and II) (modified from Doherty et al. 2011).

potential but low biological value can act as areas to "trade" development for conservation (fig. 4.5).

Clumped distributions of populations suggest that a disproportionately large proportion of breeding birds can be conserved within core areas. For example, 75 percent of the breeding population in the eastern range of sage-grouse was captured within only 30 percent of the area (Doherty et al. 2011). Wyoming is key to conservation of the species because it contains 64 percent of the known eastern breeding population and is at greatest combined risk from wind energy and oil and gas development (tables 1 and 3 in Doherty et al. 2011). Risks to core areas vary dramatically, and each state and province must do its part to ameliorate these risks to maintain sage-

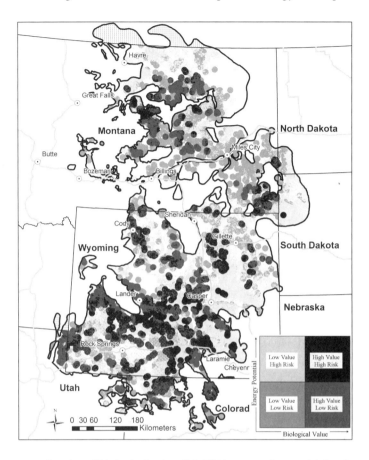

FIGURE 4.5. Overlay of biological value (25–75% core regions = high value) with energy potential for oil, gas, or wind power development to assess risk of development to greater sage-grouse core areas (modified from Doherty et al. 2011).

grouse distribution and abundance. Successful implementation of landscape conservation in one state is insufficient to compensate for losses in others.

Core areas provide a vision for decision makers to spatially prioritize conservation targets. Core area analyses and associated geo databases are now publicly available online for use in range-wide sage-grouse planning (Doherty et al. 2010b). Several western states adopted the initial concept and have subsequently refined core areas by linking them with the best available habitat maps and expert knowledge of seasonal habitat needs outside the breeding season. Core areas have been heralded as a way of partnering with industry to fund conservation in priority landscapes (box 9.2 in chap. 9) and as a basis for forecasting development scenarios to aid in conservation design (chap. 10). Identification of core areas provides a biological foundation for implementing community-based landscape conservation (chap. 12). Landscape-scale conservation in priority areas is the most defensible and realistic solution to the dilemma between energy development and sage-grouse conservation in the West. Maintaining large landscapes with minimum disturbance is paramount to sage-grouse conservation and will require collaborative efforts from a diverse group of stakeholders.

A pressing need in conservation plans is a better understanding of connectivity between sage-grouse populations (Oyler-McCance and Quinn 2011). Our understanding of sage-grouse movements, dispersal, and connectivity is limited because telemetry studies have not been conducted to document how individual populations move during dispersal or seasonal migration (Knick and Hanser 2011). Analytic advances in landscape genetics (Murphy et al. 2010) and noninvasive sampling of genetic material from feathers collected off leks may provide an efficient means of quantifying connectivity between sage-grouse populations (Storfer et al. 2007). However, genetic samples alone can obscure emerging or disrupted patterns of animal movements (Fedy et al. 2008). New GPS technology is a promising technique to identify sage-grouse movements that may be critical to population persistence but could be missed by radio telemetry and might not be detected using genetic approaches (i.e., migration).

Lastly, researchers should focus on finding links between prescriptive management actions and sage-grouse productivity. The initial requirement is an understanding of how different vital rates influence overall population growth and the plasticity and ability to manage influential vital rates. Once key vital rates are identified, management practices that bolster vital rates should be implemented to help maintain and enhance populations. Tools to manage sage-grouse populations will vary across the species range with biotic and abiotic characteristics of different landscapes and local constraints to populations. Some population may benefit from changing grazing regimes, removing conifers, or managing invasive species, yet these will ultimately depend on the site. The ultimate measure of our management success will be the biological return on investment, as measured in number of birds.

Off-Site Mitigation for Impacts Resulting from Development

Mandatory off-site mitigation for sage-grouse beyond that of voluntary compliance and the corporate mantra of sustainability may someday become a reality (chap. 9). If and when it does, biodiversity offsets could provide a mechanism to compensate for unavoidable damages from new energy development as the United States increases domestic production. To date, proponents argue that offsets provide a partial solution for funding conservation, while opponents contend the practice is flawed because offsets are negotiated without the science necessary to back up resulting decisions. Missing in negotiations is a biologically based currency for estimating sufficiency of offsets and a framework for applying proceeds to maximize conservation benefits.

One new study provides a common currency for offsets for sage-grouse by estimating the number of birds affected at levels of oil and gas development commonly permitted (Doherty et al. 2010a). Analyses used lek count data from across Wyoming ($N = 1,344$) to test for differences in rates of lek inactivity and changes in bird abundance at five intensities of energy development, including control leks with no development. Impacts are indiscernible at twelve wells per 32.2 square kilometers (1 well pad per square mile). Above this threshold lek losses were two to five times greater inside than outside development, and bird abundance at remaining leks declined by 32–77 percent (Doherty et al. 2010a). Documented impacts relative to development intensity can be used to forecast biological tradeoffs of newly proposed or ongoing developments, and when drilling is approved, anticipated bird declines form the biological currency for negotiating offsets. Implications suggest that offsetting risks using birds as the currency can be implemented immediately; monetary costs for offsets will be determined by true conservation cost to mitigate risks to other populations of equal or greater number. If this information is blended with landscape-level conservation planning, the mitigation hierarchy can be improved by steering planned developments away from conservation priorities, ensuring that compensatory mitigation projects deliver a higher return for conservation that equate to an equal number of birds in the highest-priority areas, provide on-site mitigation recommendations, and provide a biologically based cost for mitigating unavoidable impacts (chap. 9).

Conclusion

The severity of impacts on sage-grouse populations from various types of energy development dictate the need to shift from local to landscape conservation. This shift should transcend state and other political boundaries to develop and implement a plan for conservation of sage-grouse populations across the western United States and Canada. Tools are available that overlay the best-known areas for sage-grouse with the extent of current and projected development for all of Sage-Grouse Management Zones I and II (Doherty et al. 2011), and range-wide core areas for conservation planning are publicly accessible (Doherty et al. 2010b). Maps depicting locations of the largest remaining sage-grouse populations and their relative risk of loss provide decision makers with the information they need to implement community-based landscape conservation (chap. 12). Ultimately, multiple stressors—not just energy development—must be managed collectively to maintain populations over time in priority landscapes.

A scientifically defensible strategy can be constructed, but the most reliable measure of success will be long-term maintenance of robust sage-grouse populations in their natural habitats. Forgoing development in priority landscapes is the obvious approach necessary to conserve large populations. The challenge will be for governments, industries, and communities to implement solutions at a sufficiently large scale across multiple jurisdictions to meet the biological needs of sage-grouse. New best-management practices can be applied and rigorously tested in landscapes less critical to conservation. We have the capability and opportunity to reduce future losses of sage-grouse to energy development, yet populations continue to decline as energy production increases, so the need for interjurisdictional cooperation is paramount. Political wrangling, lawsuits, regulatory uncertainty, and repeated attempts to list the species as federally threatened or endangered will continue until we demonstrate success in collaborative landscape planning and on-the-ground actions that benefit sage-grouse populations.

Acknowledgments

Funding for this work came from the state offices of the U.S. Bureau of Land Management in Montana and Wyoming, Wolf Creek Charitable Foundation, the Liz Claiborne and Art Ortenberg Foundation, and the University of Montana.

Chapter 5

Effects of Energy Development on Ungulates

MARK HEBBLEWHITE

Increased energy consumption and overreliance of the United States on foreign energy has led to an increase in the development of domestic resources. This national policy manifested in western North America especially through the late 1990s and 2000s. For example, between 2002 and 2006 in Montana, oil production increased by 213 percent, the number of oil wells by 17 percent, and the number of natural gas wells by 34 percent (Montana Board of Oil and Gas Conservation 2006). Increases are similar to the nearly 60 percent increase in the number of permit applications throughout the West in the last decade (American Gas Association 2005). Although this relative growth is impressive, comparison with the heavily developed oil and gas fields of Alberta reveals that Montana production is less than 10 percent of currently active oil and gas wells in Alberta. Thus, from an energy development perspective, energy impacts on wildlife in the conterminous United States are just getting started.

Energy development can affect almost all natural resources, including surface and subsurface hydrological processes, natural disturbance regimes such as fire, soil erosion processes, wildlife habitat, and wildlife population dynamics (Naugle et al. 2004; Bayne et al. 2005a, 2005b). Limited regulatory mechanisms may be in place for a few key wildlife species, including greater sage-grouse (*Centrocercus urophasianus*; chap. 4), but mitigation is typically implemented on a site-by-site basis. Regardless of small-scale regulations often applied to individual well site permits, impacts of development are most often felt through cumulative effects of not just one well site at a time but across large landscape scales on the order of thousands of square kilometers (Kennedy 2000; Schneider et al. 2003; Aldridge et al. 2004; Johnson et al. 2005; Walker et al. 2007a; Frair et al. 2008). Thus, management agencies face the difficult task of sustaining wildlife populations at large landscape scales in the face of small-scale and piecemeal environmental impact assessment (chap. 11).

In this chapter, I review effects of energy development on large mammals, with a focus on ungulates in western North America. I emphasize ungulates because of recent interest by the public and management agencies on effects of development on these focal species. Ungulates also provide a useful entry point to understanding energy impacts on wildlife because as herbivores, they must balance risk of being killed by predators with changes in forage availability (Hebblewhite and Merrill 2008), and energy development can influence the entire food web in which ungulates live (DeCesare et al. 2009). Indeed, often the indirect effects of food web dynamics influence focal species after development (chap. 3). Objectives of this chapter are to synthesize the literature about the effects of energy development on ungulates, identify weaknesses of existing research to provide guidelines for the management of energy development, and propose a conceptual framework for understanding effects of development on ungulates. Given substantial shortcomings in the existing approaches used to study the effects of energy on large mammals, I conclude with recommendations to improve the science of energy impacts on wildlife.

Biological Response of Ungulates to Energy Development

I conducted a literature search of energy–ungulate impact studies using searches of electronic databases from 1970 to the present, including ISI Web of Science, Google Scholar, Absearch, Bioabstracts, Biological Abstracts, Environmental Sciences, Dissertation Abstracts, government resources, Geology Abstracts, and Forestry Abstracts. I searched databases using combinations of the keywords *elk, mule deer, pronghorn, woodland caribou, energy development, petroleum development, oil development, gas development, wildlife,* and *ungulate in western North America*. I recorded information on each study regarding study area, methods, results, recommendations, and implications. I found 120 publications that met search criteria. Seventy were field studies that investigated aspects of energy development on ungulates. Of those seventy studies, almost half were peer-reviewed scientific publications, 30 percent were unpublished reports, 11 percent conference proceedings, and a combination of book chapters and graduate theses made up the remainder. Elk were the most common ungulate studied in forty-five studies, followed by woodland caribou (twenty-nine), mule deer (twenty), and pronghorn (twenty-one). A surprising number of literature reviews (ten) have been conducted on this scant literature.

From a study design perspective, most studies ($N = 27$, 47 percent) used a weak observational approach in which impacts of development were inferred from correlations between levels of human activity and measures of ungulate responses to treatments. Comparative designs, where responses were evaluated before or after development, but without a control, were used in 19 percent ($N = 11$) of the studies. Only ten studies (18 percent) used the most powerful experimental design, a before–after control–impact design (BACI) (Krebs 1989; Underwood 1997). Three studies were specifically designed to be predevelopment studies conducted at or before the beginning of development (Amstrup 1978; Ihsle 1982; Sawyer et al. 2002). None of the studies was replicated. Approximately 51 percent of studies used radio telemetry, collaring more than 2,000 animals. The most common alternative methods were aerial surveys (15 percent) and pellet or sign and track surveys (20 percent). Average sample size (N) used in energy–wildlife studies was 57.5, the median 39.5, considering the sample unit as the individual animal (Otis and White 1999; Gillies et al. 2006) (table 5.1).

Size of the ungulate population affected by development averaged 3,950 animals, with a median of 1,000 (table 5.1). From a sampling perspective, then, the average telemetry-based study sampled a mean of 1.5 percent or a median of 4 percent of the population. In radio telemetry studies,

TABLE 5.1. Summary statistics for literature on the effects of energy development and human disturbance on ungulates ($N = 126$ studies).

Metric	Mean	Median	Range	SD
Sample size	57.5	39.5	4–223	53.6
Number of animals marked in telemetry studies	58.7	34.2	4–223	60
Number of telemetry locations per animal	22	17	1–55	
Population size	3,950	1,000	35–48,000	22,058
Study area size (km²)	3,882	798	26–20,000	5,924
Study duration (years)	2.7	2.1	0.15–11	2.28

only twenty-two very-high-frequency (VHF) locations were obtained per animal per study. Pseudoreplication (Hurlbert 1984) occurred in 30 percent of studies, most commonly where telemetry locations were considered the sample unit.

Oddly, studies often failed to report the size of the study area, a key parameter influencing magnitude of impacts, spatial scaling, and density of disturbances. Where size was reported (N = 56), it ranged from 26 to 190,000 square kilometers. Studies of boreal woodland caribou populations had the largest study areas, averaging 28,000 square kilometers (range 225–190,000 square kilometers), and were statistically larger than those for other species (ANOVA; P < .01). Excluding caribou, the largest study area in the lower forty-eight states was 15,000 square kilometers in Wyoming (Sawyer et al. 2005b), with no other differences between species (P > .30). Although the average size of study areas appeared large (3,382 square kilometers), the median was only 798 square kilometers, a 15-square-kilometer radius (table 5.1).

Studies were short, paralleling the duration of active energy development. Average and median duration were 2.7 and 2.1 years, respectively. Most studies were conducted in two time periods, the first in the 1980s and the second of which we are currently experiencing (hence this book; fig. 5.1).

Two peaks in the number of studies correspond closely (r = .57, P = .09) with peaks in energy exploration and development in the last 30 years (American Gas Association 2005; Montana Board of Oil and Gas Conser-

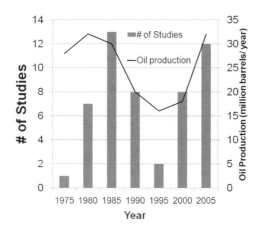

FIGURE 5.1. Frequency of study date for studies (N = 60) of the effects of energy development on ungulates plotted against peak oil production in Montana (Montana Board of Oil and Gas Conservation 2006).

vation 2006). Careful reading revealed that of just the studies designed to investigate effects of energy development activities, nearly 70 percent ($N = 56$) were reactionary and designed largely as consultancies to monitor and mitigate environmental concerns about the development as a condition of the drilling or exploration permit (e.g., Horesji 1979; Irwin and Gillian 1984; Morgantini 1985; Johnson and Wollrab 1987; van Dyke and Klein 1996).

Next, I briefly review effects of development on the main ungulate species, drawing parallels between species-specific effects. I start with woodland caribou; although they are unfamiliar to readers in the lower forty-eight states, I begin with these endangered species because more long-term and large-scale research concerning effects of energy development has involved caribou than other ungulates. Research on caribou can be explained in part by the accelerated rate of development of the boreal forest in Alberta, Canada, and because woodland caribou are sensitive to anthropogenic changes to community dynamics.

Woodland Caribou (Rangifer tarandus tarandus)

I focus on effects of development on boreal woodland caribou rather than those on barren-ground caribou (*R. t. grantii*), which have been summarized elsewhere (Cronin et al. 1998, 2000; National Research Council 2003; Johnson et al. 2005). Research on woodland caribou has progressed largely in three phases: studies on (1) initial effects of exploration; (2) altered ecosystem dynamics that influence caribou population processes (e.g., survival, growth); and (3) regional, cumulative effect assessments that address population viability at regional scales. In subsequent sections on elk and other species, I draw parallels between caribou research and ungulate–energy impacts in the lower forty-eight states, where I argue research is being conducted largely at the first or second step.

A consilience of findings across studies of the boreal forest confirms that the decline in caribou populations is attributable to large-scale changes to predator–prey dynamics as a result of forestry and energy development (Committee on the Status of Endangered Wildlife in Canada 2002; Alberta Woodland Caribou Recovery Team 2005). Historically, caribou coexisted at large spatial scales with moose and wolves. Caribou adopted a spatial separation strategy whereby they selected large contiguous tracts of habitat, such as peat bogs and old-growth conifer, that were unsuitable for wolves and moose, the wolves' primary prey (James et al. 2004). Increased forestry produced early seral stands, which provided an abundance of forage, which in turn increased moose populations. Higher wolf populations soon followed (Fuller et al. 2003), which upon exceeding a density of about seven wolves per 1,000 square kilometers exerted enough secondary predation on caribou to reduce their survival rate and drive population declines (Stuart-Smith et al. 1997; James and Stuart-Smith 2000; McLoughlin et al. 2003; Alberta Woodland Caribou Recovery Team 2005).

Energy development exacerbates impacts from forestry by adding to the landscape high densities of seismic exploration lines. Studies that examined the impacts of well site development or seismic exploration confirmed the negative impacts of exploration on caribou. This formed the basis for early regulations designed to minimize the timing of development overlap with key calving seasons and late winter seasons. In effect, this policy is a formulation of the hypothesis that the main impacts of development are behavioral only and that through avoidance of key behavioral periods, development impacts can be reduced. This policy was tested in a series of experimental and modeling studies. Bradshaw et al. (1997, 1998) showed that the negative impacts of disturbance caused by seismic exploration explosions increased caribou movement rates and habitat shifts and reduced feeding times. Behavioral changes resulted in potential loss of body mass and reduced reproduction, linking avoidance to population declines. Also, wolves travel at higher speeds on seismic lines (James et al. 2004), which increases kill rates on large ungulate prey species (Webb et al. 2008; McKenzie et al. 2009) and increases the overlap of wolves and caribou (Neufeld 2006). As a result, caribou show strong avoidance of human development near roads and seismic lines, as well as well sites (Dyer et al. 2001, 2002). Dyer et al. (2001) documented maximum caribou avoidance of areas 250 meters from roads and seismic lines and 1,000 meters from wells, which, when extrapolated to the entire study area, affected 22–48 percent of available caribou habitats with potential road avoidance effects. Dyer et al.'s (2001) results presented the first clues that human development impacts were operating cumulatively and at large spatial scales. Yet the magnitude of observed impacts in these simulation studies was less than the rate of declines of some caribou herds, suggesting the next round of studies that investigated dynamics at the level of the individual caribou herd.

During the next phase, scientists began studying population dynamics of affected caribou herds across Alberta, confirming that the majority were declining (Alberta Woodland Caribou Recovery Team 2005; McLoughlin et al. 2005) for the reasons described earlier. Empirical (McLoughlin et al. 2005) and modeling research at this stage confirmed the grim predictions of the cumulative effects of landscape change on caribou (Weclaw and Hudson 2004; Lessard et al. 2005; Sorensen et al. 2008). We now know that dramatic changes in energy development policy and aggressive measures, such as landscape restoration, core protected areas, and development restrictions, may be necessary to recover this federally threatened species (Alberta Woodland Caribou Recovery Team 2005). Unfortunately, small-scale mitigation efforts to restore seismic lines using experimental line blocking experiments failed to achieve any measurable reduction in travel by wolves. Neufeld (2006) concluded that seismic line restoration at the scale necessary to reduce predation risk on caribou was unfeasible and that large-scale mitigation is a key to conservation.

Cumulative effect assessment at large scales confirmed the grim picture facing caribou conservation in the face of energy development in Alberta. Schneider et al. (2003) developed cumulative effect assessment scenarios for caribou herds in Alberta and showed that even under optimistic scenarios in development rates, available caribou habitat would decline from 42 percent of the study area (59,000 square kilometers) at present to about 6 percent within 100 years. Empirical cumulative effect models also confirm the dire straits caribou face. Sorensen et al. (2008) compared the caribou population growth rate with the total amount of industrial development within caribou ranges and the total amount of caribou ranges burned by fire. This simple management model successfully predicted the expected caribou population growth rate as a function of percentage industrial development and percentage area burned (Sorensen et al. 2008; fig. 5.2).

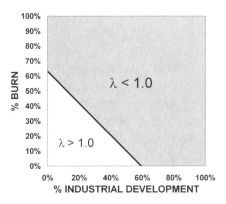

FIGURE 5.2. Meta-analysis model for woodland caribou population growth rate as a function of the percentage of the boreal caribou range that was burned and the percentage of the caribou range converted to nonhabitat through industrial development. The regression model was developed using six woodland caribou population ranges across a 20,000-square-kilometer area in northern Alberta, and is described by $\lambda = 1.191 - (0.314 * \text{IND}) - (0.291 * \text{BURN})$ ($R^2 = .96, N = 6, P = .008$) (modified from Sorensen et al. 2008).

Therefore, from a simple management perspective, the key variables after decades of research boiled down to the amount of habitat lost, which disproves the implicit policy hypothesis that energy development can be mitigated with timing or seasonal restrictions, and also refutes the hypothesis that incremental continued energy development is consistent with caribou persistence.

Today, caribou are listed as a threatened species both federally and provincially, with more than 60 percent of identified herds in Canada declining because of some form of industrial human development (Alberta Woodland Caribou Recovery Team 2005). Drastic recovery actions are being proposed, and the federal government is developing critical habitat designations that will undoubtedly result in recommendations for restricting the amount of industrial development allowed within declining caribou ranges. In summary, we have learned the following conservation lessons from the caribou–energy development story in Alberta: Short-term disturbances from energy exploration phases were not necessarily the most significant population-level impacts; by the time population-level impacts were detected, it was almost too late to recover many populations, or the level of restoration activities needed was unfeasible; it was the amount of habitat destroyed by humans, not habitat fragmentation effects per se, that caused declines; the sample size was effectively the population of caribou for statistical, biological, and planning reasons; and cumulative impacts were not always evident from individual studies, and scaling up to regional scales was needed.

Elk (Cervus elaphus)

Studies of the effects of energy development on elk have largely investigated impacts during exploration, with few studies focusing on population-level impacts and almost none examining cumulative effects. For example, van Dyke and Klein (1996) studied the effect of active drilling on elk near Line Creek Plateau in Montana by comparing seasonal and annual home range characteristics and use of cover for ten VHF-collared elk from 1988 to 1991. They compared home range size, home range centroid, and coarse-grain habitat use by elk before, during, and after development, with each phase lasting 1 year. Elk in the study site and the control site had significantly different distributions, suggesting a normal seasonal change rather than effects of drilling. Elk were rarely found outside the forest during the day while activity was taking place at the well sites. Elk responded to disturbances by shifting their use of the range, centers of activity, and use of habitat, and the authors concluded that elk do not abandon their home ranges during well site development, and they quickly return to predevelopment conditions after development. Unfortunately, limitations of this study are many, including small sample sizes; only 474 locations, ten elk, and two seasons (winter and summer) over approximately 4 years of the study yields approximately six locations per elk per season-year, which is woefully low for reliable home range and centroid estimation (Powell 2000). A second problem was scale: This study evaluated the effect of a single oil well in an approximately 500-square-kilometer area, a density of 0.003 wells per square kilometer, a trivially low density for such a large area. The utility of this study to current development, where dozens of wells are being drilled simultaneously in an existing matrix of developed oil fields, is questionable.

Other studies used radio telemetry to examine the effects of seismic exploration on elk (Johnson and Lockman 1979; Hiatt and Baker 1981; Olson 1981; Irwin and Gillian 1984; Gillin 1989; Hayden-Wing Associates 1990; van Dyke and Klein 1996). By and large, these were observational studies with poor experimental design, with few or no predevelopment data, of short duration, or with ridiculously small sample sizes (e.g., $N = 6$; Olson 1981). As an exceptional example of weak experimental design, Hiatt and Baker (1981) evaluated effects of drilling a single well on elk in Wyoming by comparing track counts in a 9-day period before development with track counts after development. Despite weak inference, results of these studies generally support the conclusion that elk move away from active exploration areas, altering their habitat selection, movement rates, and use of areas in their seasonal home ranges, but do not shift or change home ranges and merely redistribute within their home ranges.

In a unique study, effects of pipeline construction on movements of elk, moose, and deer were evaluated in west-central Alberta (Morgantini 1985). Using snow track surveys, crossing attempts through the pipeline during construction were documented for seventy-six ungulate groups. The pipeline was a barrier for 53.9 percent of ungulate groups that tried to cross them. Several practical recommendations are provided to maintain periodic openings in pipelines under construction and even underpasses or overpasses along pipeline to mitigate crossing barriers.

There were few examples of well-designed comparative or experimental studies on elk habitat selection and indirect loss of habitat from energy development. In a recent study, Sawyer et al. (2007) examined the response of elk in open habitats to distances to roads in a system with low densities of oil and gas development at present but with moderate to high development potential. Sawyer et al. developed seasonal resource selection functions (Boyce and McDonald 1999) using telemetry locations from thirty-three Global Positioning System (GPS)-collared female elk and validated them against fifty-five VHF-collared elk. Elk selected for summer habitats with higher elevations in areas of high vegetative diversity, close to shrub cover, with northerly aspects and moderate slopes, away from roads. Winter habitat selection patterns were similar, except elk shifted to areas closer to roads than in summer, indicating a strong response of road avoidance during summer. Results suggest that elk can meet their year-round needs with low traffic. Similarly, elk avoided roads and active gas and oil well sites the most during summer in the Jack Marrow Hills, Wyoming (Powell 2003), strongly selecting for habitats more than 2,000 meters from these features. Avoidance of roads and well sites declined in fall, winter, and spring, when elk avoided only areas less than 500 meters from human development. During calving (May 15–June 30), elk avoided areas less than 1,000 meters from roads and wells. These studies make the important observation that elk continued to avoid energy development long after exploration was completed, and findings open the door to examining potential population-level impacts if areas continue to be developed. Unfortunately, no studies of elk examined population-level impacts.

Pronghorn (Antilocapra americana)

Studies of energy development and pronghorn have focused less on the effects of exploration and more on disruption of migration routes, changes in habitat selection, and population-level impacts. Foci represent marked improvements over most studies on elk. Given that most pronghorn studies are quite recent, they seem to capture the same phenomenon as caribou studies in that by the time impacts are detected, populations may have already started to decline.

The series of studies by Berger and colleagues (Berger 2004; Berger et al. 2006a, 2006b, 2007) examined the response of pronghorn to energy development in the Upper Green River as a 5-year project (still ongoing), overlapping the study area of Sawyer et al. (2002). This area is underlain by the Jonah and Pinedale Anticline natural gas formations, estimated to contain more than 283 billion cubic meters of natural gas and coal bed methane deposits, and the area is undergoing rapid expansion. Energy development started here in 2001, so studies by Berger, Sawyer, and colleagues assess only early impacts of development. Goals were to investigate the effects of natural gas development on pronghorn behavior, migration, habitat selection, and, ultimately, population consequences. Study design was strong, with GPS collaring of about fifty pronghorn per year and 100 VHF collars per year in a control and energy development area.

Berger et al. (2006a, 2006b) reported that the overriding natural factor influencing distribution of pronghorn on the winter range was snow depth. Despite avoidance by some individuals, at the population level, the authors did not find that pronghorn avoided infrastructure at current levels of development. From a population perspective, they also found no difference in pronghorn survival in undeveloped and developed areas. Findings suggest that development does not influence pronghorn, but the authors caution that results are preliminary, winters have been mild during the study (impacts may be greater during harsh winters with deeper snow), the area of most intense development is not prime pronghorn habitat, and responses may be expected to increase over longer periods of time for long-lived ungulates than the 2-year time window reported on to date. Regardless of the equivocal results of energy development on pronghorn winter ranges, the studies by Berger et al. (2006a, 2006b, 2007) showed dramatic effects of development on migration at the regional scale, which we return to later in the discussion.

In a particularly illustrative example, Easterly et al. (1991) conducted a study to examine the effects of energy development on both pronghorn and mule deer in the Rattlesnake Hills of Wyoming. Their study was in response to repeated violation by the Bureau of Land Management (BLM) of the 1985 environmental impact statement on the Platte River Resource Area of their own policies regarding timing restrictions of development on crucial winter range. Despite a federal policy of no surface development between November 15 and April 30, the BLM issued eighteen permits for drilling operations in crucial winter range between 1987 and 1991. Easterly et al. tested whether violation of this policy was negatively affecting ungulates, but they collected no predevelopment data and had no controls or comparison sites. They used a combination of radio telemetry and aerial and ground surveys to measure home range responses, densities, movements, and survival as a function of human development. Pronghorn densities were substantially lower closer to energy development, and radio-marked pronghorn avoided well sites during disturbance. The prime limitation of this well-designed study was the lack of predevelopment data on mule deer and pronghorn distribution in the region.

Mule Deer (Odocoileus hemionus)

More studies have focused on reasons for population declines of mule deer, in part because of broad-scale declines in mule deer productivity across western North America (Unsworth et al. 1999; Gill 2001). Factors resulting in declines of mule deer populations in Colorado included competition with increasing elk populations, density dependence in vital rates caused by historic high population densities, long-term declines in habitat quality for mule deer because of changes in fire history regimes, overharvest, increasing predator populations, and disease, including chronic wasting disease (Gill 2001). Energy development can now be added to the list throughout much of mule deer range. Long-term (6- to 8-year) and large-scale studies (e.g., wildlife management units; about 1,000 square kilometers) are needed to rigorously assess causes for mule deer declines (Gill 2001). Their recommendations are relevant for considering the effects of energy development on large ungulates.

Early studies on mule deer paralleled those of elk in their evaluation of early phases of development. For example, Ihsle (1982) and Irby et al. (1988) worked in the same study area, conducting an observational (without a control) study over a 10-year period during oil field development on the east slope of the Rocky Mountains in Montana. Early on, development was minimal, with less than 0.003 wells per square kilometer. Phase I findings showed almost no impacts of development on mule deer home range, movement, habitat selection, migration, and fawn-to-doe ratios (Ihlse 1982). They found no effects of development because oil wells were restricted to a small part of the study, development density was very low, and the spatial scale of the study area was large. Similarly, Easterly et al. (1991) found equivocal effects of development on mule deer. Densities of mule deer were similar close to and far from drilling activities, but mule deer were located farther from development during drilling, but not after, when they were the same distance as before development. This indicates some habituation response of mule deer to development.

More recently, Sawyer et al. (2005a, 2005b, 2006, 2007) conducted a series of related studies on effects of energy development on mule deer in the Pinedale–Jonah Anticline in southwest Wyoming. Initial studies focused on migration of radio-collared mule deer ($N = 158$) and pronghorn ($N = 32$) and noted the potential for energy development impacts on migration corridors. From a habitat perspective, Sawyer et al. (2006) reported expanding development over a 5-year period with an increase of 95 kilometers of roads, 324 hectares of well pads, and a total of about 400 hectares of lands directly lost to development footprints in the study area, an increase in density of 0.12 kilometers per square kilometer and about 0.3 wells per square kilometer. Mule deer avoided areas close to development, responses to development occurred rapidly (within 1 year of development), and avoidance of development increased over the course of the 3-year study. Sawyer et al. (2006) reported lower predicted probabilities of use within 2.7–3.7 kilometers of well sites, confirming that indirect habitat losses far exceeded direct losses. Over the study, areas classified as high-quality habitat before development changed to low quality, and vice versa, showing that mule deer shifted their habitat use away from high-quality habitats to marginal habitats in response to development. Presumably, such responses will have population implications, but Sawyer et al. (2006) did not examine them. The authors recommend demographic studies and activities that reduce the footprint associated with development, including directional drilling from single well pads to multiple gas sources to reduce surface impact, limited public access, road networks developed with the goal of minimizing new road construction, and guidelines to minimize human disturbance during winter and on designated high-quality ranges.

In a related study, Sawyer et al. (2005b, 2006) focused on predevelopment phase mule deer ecology from 1988 to 1991 before development started, from 2001 to the present, in the Sublette mule deer herd near Pinedale, Wyoming. With the preliminary data collected in Phase I and two treatment areas in Phase II (with and without development), this study represents a well-designed BACI study. Before development, the Sublette mule deer population was a healthy and productive population, with adult female survival rates (0.85, $N = 14$) and a fawn-to-doe ratio (more than 75:100) indicative of a growing population (Unsworth et al. 1999). In 2002, mule deer densities were similar between the control and energy development treatments, but they have been diverging since 2002. In the developed area, mule deer densities declined by about 47 percent over a 4-year period ending in 2005, whereas in the control area, there was no negative trend, and mule deer densities were constant and similar to predevelopment density on the treatment area. This suggests a demographic impact of energy development, yet survival differences in adult female and overwinter fawn survival were not statistically different between the two areas. Sawyer et al. (2005b) speculate that they found no demographic difference between treatments because small-scale demographic differences could explain the differences in population trend, but they are preliminary and influenced more by small sample size and will be verified later by more detailed analysis; or differences were driven by emigration or dispersal from the developed areas. Migration routes were also identified, as discussed later, in Phase I. Results echo conclusions from caribou and pronghorn studies that the effects of energy development often take a long time to manifest on ungulate populations, if present, and detecting these effects is the biggest challenge.

Discussion

Readers who had hoped that a clear picture would emerge about how to mitigate effects of energy development on ungulates are probably disappointed, and this is perhaps the most important message from this chapter. Previous reviews provide strategies for mitigating small-scale effects of disturbance on ungulate behavior, yet most conclude by admonishing managers to conduct more long-term, population-based studies. Unfortunately, my conclusions from reviewing the literature are that, at least for ungulates, few have heeded this advice. During the current energy development rush, sadly, there are still few clear evidence-based management recommendations that will definitively mitigate the impacts of energy development on ungulate populations (emphasis on populations).

A second major conclusion is that energy development studies proceed in the following manner (sensu Lustig 2002): (1) A well drilling permit is applied for on an ungulate winter range; (2) the permit is granted with stipulations that attempt to reduce impacts by applying timing restrictions at critical life stages (e.g., calving); (3) either because stipulations are knowingly violated or as an additional stipulation, a study is commissioned to investigate effects of development on ungulates; and (4) the "monitoring" study is often designed hastily, with inadequate resources, sample size, temporal or spatial scope, and experimental design such as predevelopment data, and no commitment to monitoring beyond the intended life of the development phase. Thus, I conclude that wildlife biologists, as a profession, are failing to live up to professional standards and guidelines of the Wildlife Society by agreeing to participate in poorly designed studies that are aimed merely at appeasing the small-scale regulatory process. The large number of animals captured and handled (more than 2,000), their capture-related mortality, the financial investments made by energy companies, and investments in personnel time do not weigh favorably against the meager knowledge base available on the effects of development on ungulate populations. Figure 5.1 reinforces the impression that most studies of wildlife–energy relationships have been reactive, driven by trends in energy production, and are not part of any proactive adaptive management program (chap. 12). I hope this review convinces some of the need for better-designed studies of energy–wildlife impacts.

I draw these conclusions for the following main reasons. First, to date, there has not been one rigorously conducted study (e.g., a replicated experiment) of the effects of energy development on ungulates with a sufficient duration of both study and energy impact to be able to draw firm conclusions about the population impacts of development on ungulates. The average duration of studies was very short (2.5 years) when compared with the lifespan of ungulates that may live for more than 20 years. Few studies actually measured adult female survival, and not one study reported effects of energy development on population growth rate for pronghorn, mule deer, or elk (caribou are the exception). Studies that did measure adult female survival failed to show any impact of energy development by and large and were conducted only for a short time period, consistent with effects of low statistical power due to small sample sizes (Gerrodette 1987) and for species with high and constant adult survival rates (Gaillard et al. 2000). For long-lived species such as ungulates, impacts of changes in the environment may take decades to manifest because of compensatory reproduction and resilience in the adult age cohort and because ungulates possess high and constant adult survival (Albon et al. 2000; Festa-Bianchet et al. 2003; Gordon et al. 2004; Coulson et al. 2005). Following from Gaillard et al. (2000) and Eberhardt (2002), energy impacts would be expected to manifest first on the least sensitive but most variable population vital rates, such as calf survival and recruitment, not the most important but least variable adult survival rates, such as adult female survival. In fact, ungulate life history in general makes it extremely difficult to determine the effects of development on populations in a 2- to 3-year study. Recent recommendations of reviews of ungulate demography studies suggest that a minimum of fifty marked adult female ungulates monitored over at least a 5-year period (Gordon et al. 2004) are needed to gain a mechanistic understanding of changes in adult survival rates linked to environmental changes such as energy development. Although population-level surveys are capable of identifying important changes (Sawyer et al. 2005a, 2005b), without detailed demographic data, mechanisms driving changes will be cause for speculation. Thus, long-term changes in the way in which agencies and industry engage in research on energy impacts on wildlife need to occur to achieve an evidence-based framework for mitigating development.

The second major reason why I conclude that impacts are poorly understood is that most studies focus only on early phases of development.

Effects of development may take years to manifest on long-lived ungulates, yet most studies were conducted either before or during the first 1–5 years of development. Short-term studies do not give populations enough years to equilibrate to development and loss of habitat. Time lags should be expected as normal and are likely to be at least one generation time for long-lived ungulate species (Gaillard et al. 2000). A major additional problem with studying impacts of development only early during development is that density of development is confounded with duration of development, again confusing clear cause–effect relationships because of the period of equilibration needed for long-lived ungulates. In an extreme example, van Dyke and Klein (1996) investigated impacts of the first oil well constructed in a nearly undeveloped area on elk behavior in hopes of estimating population-level impacts. At such low development densities, population-level responses for a large ungulate are not expected to occur because ungulates can habituate to responses at low development thresholds.

This review does provide some conclusions about behavior-level impacts of energy development on ungulate species that will be useful to planners at the level of the individual well pad or road. Many of these behavior-level impacts were already summarized in previous reviews (Bromley 1985; Girard and Stotts 1986; Hayden-Wing Associates 1991; National Research Council 2003). However, the real question is whether such small-scale mitigations, referred to as "death by a thousand cuts" (Lustig 2002), are useful to scale up to population-level responses.

At the small scale, most ungulates displayed behavioral responses that weakly to strongly avoided energy development activities during the development phase (exploratory seismic blasting, road construction, mining construction, forest operations, and well drilling). Pronghorn, elk, and mule deer, in that order, generally showed the strongest avoidance of development during the construction phase. Seasonal impacts were variable and occurred year-round in winter ranges, calving ranges, migratory corridors, and summer ranges. Early studies focused on the effects of development on winter ranges, and restrictions on crucial winter ranges are still enforced as small-scale mitigation measures to reduce impacts. However, recent studies show increasing effects of development on spring calving ranges, during summer, and especially in migration corridors. This may reflect a growing understanding of the importance of summer nutrition to ungulate demography (Cook et al. 2004; Parker et al. 2005). Regardless, recommendations for timing restrictions on spring calving ranges and critical winter ranges were echoed by a majority of studies for all species, especially elk, mule deer, and pronghorn. Unfortunately, there is little evidence that such small-scale mitigation is sufficient to mitigate effects of development at large scales. In the case of caribou, for example, we now realize that small-scale mitigation did not prevent declines resulting from large-scale cumulative impacts.

Despite this purposefully scathing critique, I draw some conclusions about impacts of development on large mammals, including the negative effects of roads, density of development, and the role of migratory movements in assessing the scale of impacts.

Effects of Roads

Roads are one of the most pervasive impacts of human development on natural landscapes (Forman and Alexander 1998), and by far their greatest impact lies in the indirect effects of habitat fragmentation and avoidance by wildlife. Current estimates indicate that the lower continental United States has about 10–20 percent of habitats affected by roads. Impacts are typically most severe near the road and extend out a variable distance, depending on the species of interest (Forman and Alexander 1998). Here I summarize the distances to which impacts extend from developments (e.g., roads, well sites). Readers should note that this zone of influence around roads does not imply 100 percent avoidance (Schneider et al. 2003; Harron 2007), yet from the information presented in studies, actual effective reductions in habitat use were not presented. For example, Dyer et al. (2001) reported on average a 40 percent reduction within 100 meters of a seismic line and declines up to 250 meters away. Powell (2003) reported 73 percent reductions in use within 2,000 meters of energy development, but other studies did not usually present enough information. In the eight studies that did report avoidance of roads, the average zone of influence extended about 1,000 meters from both roads and wells, although responses varied within seasons and between species (table 5.2).

In general, ungulates avoided roads more in summer than winter, when snow depth constrained animal movements away from roads. Regardless, even considering an effective loss of habitat of 50 percent within this zone of avoidance and a modest buffer size of 500 meters, 10 percent of a study area can be effectively lost due to indirect avoidance of roads. The effect of overlap between well sites and roads on habitat loss due to avoidance is important and deserves further investigation (Rowland et al. 2000; Frair 2005).

TABLE 5.2. Summary of ungulate studies showing avoidance of roads and well sites, with results averaged across seasons and habitat types.

		Avoidance Buffer (m)	
Author	Species	Roads	Wells
Powell (2003)	Elk	2,000	2,000
Ward (1986)	Elk	2,000	
Gillin (1989)	Elk	1,200	500
Edge (1982)	Elk	500	1,000
Rost and Bailey (1979)	Elk	200	
Frair (2005)	Elk	200	
Sawyer et al. (2005b)	Mule deer	2,700	
Dyer et al. (2001)	Caribou	250	1,000
	Average	1,131	1,125

Density of Development

I extracted density of oil and gas infrastructure (e.g., roads, wells, seismic lines) where possible, but only 17 percent of studies (twelve of seventy) that investigated direct impacts presented sufficient information (table 5.3).

Existing time-stamped datasets provide the ability to estimate densities for use in meta-analyses. I present results of a univariate meta-analysis of density of infrastructure for the twelve studies that reported an effect of development on some response variable against studies with no effect. Caveats of this simple analysis are many, and variables that could not be accounted for include size of study area, length of study, and sample size and its associated variance. Regardless, studies that reported an impact of development had higher densities of wells and roads. Impacts started to manifest on ungulate species including mule deer, pronghorn, and elk from 0.1–0.4 wells per square kilometer and 0.18–1.05 linear kilometers of roads per square kilometer. However, replicated studies are necessary to disentangle the effects of sample size, study duration, and severity and type of biological response (i.e., avoidance versus population impacts) to density of development.

Migration and Identifying Appropriate Scales

A difficult problem in ecology is how to scale up from short-term and small-scale behavioral decisions of animals to long-term landscape-scale

TABLE 5.3. Summary of density of energy development disturbance in terms of density of active well sites and linear kilometers of pipelines, seismic lines, and roads from studies where such information was reported. Despite small sample sizes of studies that reported densities, ambiguities in definition of study areas, and simplification of impacts to a binary variable, densities of disturbance appear to be related to the impact of energy development.

Study	Density of Wells (per km²)	Linear Kilometers of Roads/ Pipelines/Seismic (per km²)	Significant Impact?[a]
Sawyer et al. (2005a, 2005b)[b]	1.01	1.36	Yes
Frair (2005), Frair et al. (2008)	0.20	1.6	Yes
Sawyer et al. (2002)	n/a	0.62	Yes
Rowland et al. (2000)	n/a	0.62	Yes
Easterly et al. (1991)	0.27	N/A	Yes
Berger et al. (2006a, 2007)[b]	0.25	0.20	No
Bennington et al. (1982)	0.20	N/A	No
Olson (1981)	n/a	0.15	No
Knight et al. (1981)	0.088	n/a	No
Ihsle (1982)	0.003	N/A	No
van Dyke and Klein (1996)	<0.001	N/A	No
Summary Statistics	Mean (N)	Mean (N)	
Significant impact: yes	0.49 (3)	1.05 (4)	
Significant impact: no	0.10 (4)	0.18 (2)	

[a]Significant impact is a simple binary variable confirming whether statistically significant effects of energy development were detected on key response variables.
[b]These two sets of studies occurred in approximately the same area but defined different study area sizes based on species life history.

population responses. In the case of energy development, the question of scale also touches on the growing consensus that energy development negatively affects ungulate migration and large-scale processes. The difficulty in scaling up is why so few of the studies that showed short-term responses were able to measure or demonstrate these long-term or population-level responses. A second scaling problem is presented by Berger et al. (2007) when discussing issues of spatial scale and habitat fragmentation, both of which are totally dependent on each other (Dale et al. 2000; Turner et al. 2001). Quantifying habitat fragmentation metrics will be determined completely by the study area size, and for this reason many authors recommend conducting multiscale analyses of the effects of habitat fragmentation on wildlife species (Harrison and Bruna 1999; Turner et al. 2001).

In many of the studies I reviewed, there was a third scaling problem: that of extrapolating responses. This occurred where the effects of a local point source disturbance (well pad) were assessed at the population or home range scale, and results were extrapolated well beyond the development densities under which the response was studied. For example, van Dyke and Klein (1996) document weak or no responses of elk to the installation of a single well in an undeveloped grassland ecosystem in north-central Montana. Results of this study have been extrapolated to other wells across Montana, yet the validity of extrapolating the finding of no significant impacts to areas with higher well densities is questionable. This emphasizes the need to establish thresholds for development or broad, regional-scale cumulative impact assessments as the density of well sites and development increases.

Finally, there was often a mismatch between the spatial scale of the study in question and the spatial scale of the population under investigation that links to impacts on migration. Assuming that the goal of an impact study is to assess the impacts of a particular development on a population, unless the study area represents the annual range occupied by the ungulate population, it will be difficult to evaluate whether the changes in the population are occurring because of energy development on the winter range or because of changes occurring elsewhere in the population's range. One potential solution to the issue of how to determine the appropriate study scale is to use the spatial scale of migration as a guideline in migratory populations. Berger (2004) reviews long-distance migration throughout western North America and worldwide. Although not all populations are migratory, the reported degree of partial migration ranged from 45 to 100 percent; and most populations in studies reviewed in this article contained some migrants. Considering the one-way migration distances (35–177 kilometers across species) as a buffer suggests that the correct spatial scale to consider in evaluating the effects of energy development could range from 5,041 square kilometers for mule deer, 5,041 square kilometers for caribou, 8,464 square kilometers for elk, to nearly 19,000 square kilometers for pronghorn (Berger 2004). Guidelines suggest much larger study area sizes than are currently used to evaluate impacts. Moreover, because one of the areas where we have seen a convergence across studies is the effects of energy development on migration, studying population impacts at the migratory scale will be critical.

For example, Berger (2004) found that in the Greater Yellowstone Ecosystem about 75 percent of all large ungulate migrations have been lost due to human development. Berger (2004) illustrates the problem with a case study involving a long-term pronghorn study in the Pinedale area of Wyoming. Both residential development and future potential energy development threaten one specific migratory corridor pinch point, the Trapper's Point bottleneck, where the migration corridor narrows to less than 800 meters. In a follow-up study to this review, Berger et al. (2006b) confirm that this particular migration corridor, from the Upper Green River to Teton National Park, has probably been used for more than 6,000 years. Sawyer et al. (2009) used GPS collars on mule deer in Wyoming to monitor migration routes between winter and summer ranges in the face of impending energy development. Unlike the simpler example where Berger (2004) showed an entire pronghorn population moving through a single corridor, Sawyer et al.'s mule deer study shows that migration routes often are varied and reticulate, making protecting migratory routes challenging, and these results have been echoed for both elk and woodland caribou (Hebblewhite et al. 2006; Saher and Schmiegelow 2006). Indeed, considering the future effects of climate change, ensuring retention of migratory behaviors in the landscape may be an effective mitigation strategy. For example, a recent molecular ecology study of woodland caribou revealed important links between energy development and potential responses to future climate change. Woodland caribou in the Canadian Rocky Mountains were a mix of boreal and barren-ground caribou, and caribou with barren-ground haplotypes had a higher probability of migrating but also a higher risk of mortality because of changes to the landscape induced by energy development. Thus, energy development may be reducing migration, and in the future migratory behavior will undoubtedly help species respond to climate changes, as caribou did during the Pleistocene interglacial. Regardless, only if we create large-scale migration corridors that are protected from development or managed specifically to mitigate energy development will long-term migration persist, a critical ecological process that is declining across the Rocky Mountain West. Fortunately, the Western Governors' Association and other government agencies have recently recognized the crucial role migratory corridors play as natural mitigation because migration enables ungulates to use seasonal resources over a much larger area. Support from political bodies will aid decision-making processes to ensure protection of migratory routes.

A Conceptual Approach for Understanding Cumulative Effects of Development

I conclude here that the effects of development on ungulates are manifested through changes in the ecological communities of species, including humans, in which ungulates exist. Therefore, impacts on populations can be classified as direct and indirect impacts. Distinguishing between these two types of effects and between species is critical to identifying mechanisms and providing effective mitigation strategies. Direct effects between species (e.g., humans, development, and elk) occur when there are no intermediary species between two interacting species, for example, through direct mortality associated with energy development (e.g., roadkills, poaching; Estes et al. 2004). Most direct effects are attributed to predation or to habitat loss, such as when a population responds negatively to a reduction in available forage where development has denuded vegetation. In contrast, indirect effects occur when impacts on a species are mediated by an intermediate species. As an example, consider the indirect effect of development on sage-grouse and kit foxes (*Vulpes macrotis*) mediated by human-induced changes in avian or mammalian predators. An increase in the number of perches available to raptors indirectly increased predation rates on breeding and nesting sage-grouse (Fletcher et al. 2003; Aldridge and Boyce 2007). Similarly, coyote populations increased after development because altered landscapes support higher densities of small mammals, causing increased predation by coyotes (*Canis latrans*) on kit foxes (Haight et al. 2002). In this case, predation is proximate to the ultimate cause of human-induced changes to landscape function. Apparent competition will be a common indirect effect of human disruption of ecosystem dynamics (DeCesare et al. 2009). Therefore, effects of energy development will probably go far beyond direct impacts based purely on community ecology (Estes et al. 2004). Recent reviews have reminded ecologists that direct effects are but a fraction of the possible interactions between species in even a simple food web (Estes et al. 2004; Bascompte et al. 2005). Indirect effects of energy development also may arise because of behavioral changes by ungulates in response to energy development, such as avoidance of roads and well sites. Such findings have been corroborated across systems and at larger scales in ungulates, confirming the importance of indirect behavioral effects, such as the avoidance of predation risk and human disturbance on ecosystem dynamics (Rothley 2001; Fortin et al. 2005; Hebblewhite et al. 2005b).

Despite theoretical support for indirect effects, a cursory review of the literature reveals a myopic focus of mitigation strategies to reduce direct effects such as road mortality and habitat loss (Bureau of Land Management 2003a, 2003b). A renewed focus on the indirect effects of development mediated by community-level changes across species would provide a more complete understanding of the cumulative impacts of development.

Conclusion

I provide five recommendations regarding the impacts of development on ungulate populations.

First, current management policies make two untested assumptions about the effects of energy development on wildlife. One is that policies assume that negative impacts can be mitigated through small-scale stipulations that regulate timing and duration but not the amount of development activity. Policies also assume that wildlife populations can withstand continued, incremental development. Neither assumption is supported or refuted by evidence. Adaptive experiments are needed to explicitly test these assumptions.

Second, little scientific evidence exists to suggest that energy development will have population-level impacts on pronghorn, mule deer, or elk because rigorous and properly designed experiments have not been conducted. Instead, a host of observational studies on small-scale and short-term responses provides limited guidance to managers in search of the crucial question of population impacts. Although it is theoretically justified, relying on the precautionary principle to restrict energy development will probably be unsuccessful as an energy policy.

Third, efforts to mitigate short-term and small-scale impacts of energy development have been well described in previous reviews, albeit most often as poorly designed observational studies with weak inference. Ungulates predictably avoid areas during exploration and drilling, moving to denser cover and to areas farther from human activity. Across studies, ungulates avoided development to an average of 1,000 meters. Recommendations from previous studies still hold, namely the continued application of timing and seasonal restrictions for critical habitats and resources. However, it is increasingly apparent that small-scale mitigation alone cannot offset impacts of large-scale development on ungulates.

Fourth, scaling up from small-scale and short-term studies to population-level impacts will be difficult. One difficulty is scaling up responses of ungulates at low development densities to high densities observed in most oil and gas fields today. Preliminary analyses suggest that thresholds of development will occur at densities of 0.1–0.5 wells per square kilometer and 0.2–1.0 linear kilometer of roads and other linear developments per square kilometer. However, these results are preliminary, and more formal meta-analyses are needed. Future studies should use large-scale approaches to test for thresholds of energy development and to otherwise replicate and extend for other species what has been learned about the population viability of caribou.

Finally, an adaptive management experiment should be implemented to test whether the current energy policy provides for sustainable wildlife populations. The de facto energy policy contains untested assumptions that, if invalid, will severely affect wildlife, but no serious alternatives have been put forward as tested and proven. Alternative development policies are sorely needed if other ungulate species are to avoid the same bleak outlook as caribou in Alberta.

Chapter 6

Effects of Energy Development on Songbirds

Erin M. Bayne and Brenda C. Dale

The desire to drive our cars, heat our homes, and run our computers while still having clean air, water, and healthy ecosystems is probably one of society's greatest challenges (National Petroleum Council 2007). Debate surrounding the inherent tradeoffs of energy policy and maintaining or improving ecological health is in its infancy. The need for such discourse is urgent, however, as energy demands are increasingly resulting in development of frontier areas (i.e., Canada's boreal forest and tundra) and the last remnants of native vegetation (i.e., prairie). Informed decisions from policymakers, industry leaders, and society as a whole depend on a full understanding of how energy development affects wildlife populations (Schneider et al. 2003).

Most wildlife research related to energy impacts focuses on charismatic wildlife species, such as greater sage-grouse (*Centrocercus urophasianus*; chap. 4) and woodland caribou (*Rangifer tarandus tarandus*; chap. 5). A broader suite of wildlife indicators is needed to fully elucidate the consequences of energy development. Passerines (songbirds) are useful indicators of ecological health (Bradford et al. 1999; O'Connell et al. 2000) because they are easy to monitor and can be effective tools in garnering public support for conservation (Carson 1962). Research on forestry and agricultural practices has shown that songbirds react strongly and quickly to changes in vegetation composition, structure, and landscape patterns caused by human land use. Despite these attributes, little attention has been paid to songbirds in the debate over energy development and wildlife conservation.

In this chapter, we review studies on oil and natural gas development, and to a lesser degree coal, on songbirds in boreal forests and grasslands in western North America. Specifically, we describe mortality risk caused by energy infrastructure, the biological response of songbirds to habitat conversion and associated fragmentation, and effects of noise from energy development on songbird ecology. By outlining what is known about each of these issues relative to songbird behavior and vital rates, we highlight the critical need for more complete evaluations of the cumulative effects of energy development on songbird populations and communities at relevant temporal and spatial scales.

Bird Mortality and Energy Infrastructure

Lands developed for energy production typically include roads, communication towers, power lines, well pads, flare stacks, holding ponds, and compressor stations. The ability to fly makes birds susceptible to colliding with these features, thereby increasing mortality. Although many studies have documented mortality events at such features, there have been no attempts to determine whether this mortality is additive or compensatory to natural mortality and little effort to determine associated impacts on population dynamics (Desholm 2009).

Height of human-made structures is a well-established factor influencing the risk of collision (Mabey and Paul 2007). Television towers, cell phone towers, power poles, wind turbines, and aerials for remote monitoring equipment all have been struck by songbirds (California Energy Commission 2009). In general, songbirds are most likely to collide with such features during migration, particularly on nights with inclement weather or clouds. Mass mortality events of thousands of individuals have been regularly reported at tall (more than 200 meters) cell and television towers. Although they are not directly a part of energy sector infrastructure, the development of oil and natural gas fields and wind power facilities often results in construction of cell phone towers to facilitate communication between energy sector employees. Towers used at well pads for remote monitoring equipment, and power poles and power lines used to run equipment at energy sector sites, are typically shorter than cell phone towers (15–60 meters) and therefore should have lower collision rates. However, the larger number and broader distribution of short vertical structures may add up to a cumulative impact that is potentially greater than the more obvious mass collision events at tall towers. Despite documentation of mortality events, our understanding of the cumulative effects of vertical human infrastructure on songbird mortality, and whether it strongly influences population dynamics, remains poor.

Even less is known about collisions with structures during everyday movements by songbirds during the breeding or wintering season. Songbirds tend to be active during the day in these periods, so it is assumed that collisions are rare (Avian Power Line Interaction Committee 1994). However, methods for estimating songbird mortality due to collisions are poorly developed relative to larger birds, so current estimates may not reflect the magnitude of the problem. Songbirds have been observed to collide with power lines during the breeding season, and electrocutions have occurred when flocks of songbirds create arcs between individuals in the flock and the phase conductor (Bevanger and Thingstad 1988).

Collisions with vehicles using roads could also be a significant source of anthropogenic mortality for songbirds during the breeding and wintering season (Forman and Alexander 1998). In a reintroduction experiment with eastern bluebirds in Florida, one of the largest sources of mortality for juveniles was vehicle collisions, which played a role in limiting species recovery (Lloyd et al. 2009). Higgins et al. (2007) attributed 17 percent of observed avian mortality on a wind farm to traffic. Little is known about whether reduced traffic volume or speed reduces avian mortality. Reducing speed cannot reduce risk for wildlife unless the organism is visible at 25 meters (Ramp et al. 2006). In a forested region in Alberta, mortality of birds was higher on lower traffic volume roads with a lower speed limit, probably because surrounding vegetation was close to the road (Clevenger et al. 2003). Management of roadkills is difficult because reducing collisions may require manipulation of ditch width, vegetation height, or sound to discourage bird crossing. Reducing crossing rates may decrease connectivity between other habitats and populations.

Most energy sector companies cut hay or shrubs near energy roads and trails to reduce vehicle-induced fire risk, increase visibility, and limit snow accumulation. Haying during the breeding season destroys nests, eggs, and young of ground-nesting grassland birds and leads to increased abandonment and predation (Bollinger et al. 1990). Active aboveground nests are almost always destroyed when vegetation is cut (Frawley 1989). Removing this direct mortality is as simple as delaying haying or shrub removal to after the breeding season, but education is needed (Bollinger et al. 1990; Dale et al. 1997; Nocera et al. 2005).

Open pits where waste fluids from hydrocarbon production are stored pose a significant mortality threat to birds in some situations (Trail 2006). In 2009, more than 500 ducks were killed in a tailings pond in the tar sands of boreal Alberta. Whether waterfowl are the species most affected by open pits is debatable; the U.S. Fish and Wildlife Service identified the remains of 172 avian species (predominantly songbirds) in sampled holding ponds (Trail 2006). Dealing with other toxic byproducts can also have negative impacts. Natural gas facilities often burn off gases, such as hydrogen sulfide, during extraction. Approximately 3,000 individuals of at least twenty-six species of songbirds were found dead within 75 meters of a 104-meter flare stack in boreal Alberta (Bjorge 1987). The presence of pulmonary congestion suggested that some deaths were related to stack emissions, and other birds probably struck the stack. Even though energy sector infrastructure does result in songbird mortalities, the importance of these deaths to population dynamics remains unknown.

Dissection and Perforation of the Boreal Frontier

The boreal forest has been called the bird nursery of North America, with at least 300 species and an estimated 5 billion breeding individuals using this forest each year (Blancher and Wells 2005). This ecosystem remains one of the last unexploited wildernesses on the planet. This status is rapidly changing as a result of new energy and forestry policies in western Canada. Migratory passerines make up most of the bird species in the boreal forest. Broadly speaking, boreal passerines are split into those that prefer coniferous over deciduous forests, early versus late succession forest habitats, and upland versus lowland areas. Southern boreal forests in western Canada are dominated by deciduous tree species, whereas conifers dominate in the north. Forests do not regularly reach more than 150 years old because of frequent disturbances such as fire and insect outbreaks. Small lakes and rivers are common, and riparian habitats typically have the highest bird diversity (Hannon et al. 2002). Recent development of natural gas and non-conventional oil supplies has turned the boreal forest of western Canada into one of the largest producers of oil and natural gas in the world (chap. 2). Despite the value of the boreal forest to wildlife and the energy sector, surprisingly few studies have examined the biological response of songbirds to development.

The Case of the Ovenbird

One species for which much is known about its response to energy development is the ovenbird (*Seiurus aurocapilla*). The ovenbird is a neotropical migrant that nests and forages on the ground in mature deciduous and mixed wood forests. The ovenbird is quite common in the boreal forest of western Canada and is widely studied because of its perceived sensitivity to habitat fragmentation (Van Horn and Donovan 1994). Radio telemetry studies examining the space use of the ovenbird indicate that they do not incorporate conventional seismic lines (8–10 meters wide) in their territories, defending territories exclusively on one side (Bayne et al. 2005a; fig. 6.1; left panel).

Seismic lines are access routes typically cut by bulldozers and are the most extensive energy disturbance in the boreal forest of western Canada, averaging 1.5 kilometers per square kilometer in 98 percent of townships in northeast Alberta. Seismic line densities in some townships (100-square-kilometer areas) are as high as 10 kilometers per square kilometer (Lee and Boutin 2006). In the boreal forest of Alberta this translates into hundreds of thousands of hectares of disturbed forest. Less detailed data collected using point counts demonstrate that ovenbirds are never detected singing on pipelines, power lines, roads, or small clearings (i.e., 0.5- to 2-hectare well pads) (Bayne et al. 2008). This suggests that early successional habitats created by energy development are unsuitable for this species. This is not unexpected based on ovenbird response to forest age and its response to forest harvesting. Two fundamental questions remain despite avoidance of energy sector disturbances: Does altered behavior translate into reduced numbers of ovenbirds, and how long will such effects last given forest succession?

Given that ovenbirds avoid sites disturbed by energy sector development and an increasing area of the boreal forest is being disturbed by energy sector activity, it seems logical that fewer ovenbirds will be present in highly disturbed landscapes. However, for this to be true requires that all available habitats be used before disturbance. If available habitat is not fully occupied, then birds may simply move into remaining habitat areas. It also requires that birds be territorial and not alter their behavior by altering overlap between individuals or by reducing territory size. Bayne et al. (2005a) conducted a quantitative experiment comparing ovenbird density via spot mapping in 12-hectare stands of mature trembling aspen (*Populus tremuloides*) with no conventional seismic lines to stands with one or more seismic lines in two locations in boreal Alberta. Machtans (2006) used the same protocols to count ovenbirds in a before–after control–impact

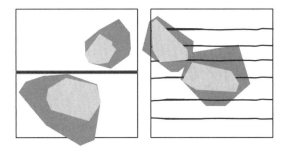

FIGURE 6.1. (Left) Example of ovenbird territories (light gray is 95% minimum convex polygon [MCP] of singing locations) and home ranges (dark gray is 95% MCP of all locations) derived from radio-marked individuals living near an 8-meter-wide seismic line cut by a bulldozer. The grid represents a series of 50- × 50-meter areas on a spot-mapping grid.
(Right) Example of ovenbird territories (light gray is 95% minimum convex polygon [MCP] of singing locations) and home ranges (dark gray is 95% MCP of all locations) derived from radio-marked individuals living in a 3-D seismic grid. The black lines are 3-meter-wide low-impact seismic lines. The grid represents a series of 50- × 50-meter areas on a spot-mapping grid.

(BACI) study in mixed-wood forests of the Northwest Territories. In two of the three areas, ovenbird density declined with increasing seismic line density, and in the third there was no change. This suggests that the behavioral avoidance of seismic lines can lead to local reductions in population size.

The energy sector has been working to narrow seismic lines in an effort to mitigate their impacts. Recent advances in seismic technology allow the use of seismic lines no more than 3 meters wide for exploration (fig. 6.1; right panel).

Ovenbirds include such lines in their territories, and no change in ovenbird density occurs relative to forest interiors (Bayne et al. 2005b). However, decreased line width has come at a cost because more lines are being cut than ever before. An increasing need for detailed seismic information has led to lines cut on tighter spacing (every 30–100 meters instead of the original 1–2 kilometers) in what is known as 3-D seismic. Whether birds react to narrow seismic lines the same way in all forest types is unknown. In mature to old forest these narrow seismic lines become difficult to locate within a few years, but in a young stand with higher tree density, the lines can be visually obvious for extended periods of time.

Representativeness of the Ovenbird

Whether results from the ovenbird can be generalized to a large number of other bird species is unclear. The ovenbird prefers forest between 60 and 100 years of age. Other forest bird species reach their highest densities in forests that are far older, suggesting that early successional habitat created by the energy sector is probably avoided by these species as well (Song 2002). However, many old-growth forest birds use the shrub or canopy layer for nesting and foraging. Whether gaps created by seismic lines act as territory boundaries for species that function in these habitat strata cannot be assumed based on behavior of the ground-dwelling ovenbird.

Based on spot-mapping data, Machtans (2006) found no differences in density of shrub- or canopy-nesting birds near conventional seismic lines. Individuals from fourteen species were observed singing on both sides of the line, suggesting that they may have included seismic lines in their territories. This may occur in part because the canopy often grows over conventional seismic lines quickly, whereas regeneration on the ground may take longer (Lee and Boutin 2006). Recent research in the Northwest Territories indicates that it takes 30–40 years for conventional seismic lines to have sufficient vegetation regrowth at ground level for ovenbirds to regularly include seismic lines in their territories (Hedwig Lankau, University of Alberta, unpublished data).

Seismic lines are not the only type of linear feature created by the energy sector. Pipelines, power lines, and roads all dissect the boreal forest and are becoming an increasingly large component of the landscape as energy fields are developed (Schneider et al. 2003). Results from the ovenbird suggest differential responses to linear feature width. Similar results have been found in eastern North America in response to low-use roads and power lines (Rich et al. 1994). Thus, best practices used by the energy sector to make features smaller and regenerate faster will create less change in bird communities. Unfortunately, no research has been done on the territorial behavior of boreal songbirds near wider linear features, such as pipelines, power lines, or roads, to identify critical thresholds in widths of features that birds will use. Point counts and nest searches on pipelines more than 15 meters wide demonstrate that forest specialists are rarely detected singing on lines, and few nests are located there (Fleming and Schmiegelow 2003). Conversely, generalist birds such as the American robin (*Turdus migratorius*), chipping sparrow (*Spizella passerina*), Lincoln's sparrow (*Melospiza lincolnii*), Le Conte's sparrow (*Ammodramus leconteii*), swamp sparrow (*Melospiza georgiana*), and common yellowthroat

(*Geothlypis trichas*) that prefer early successional habitat are more likely to be found on or near wide linear features (Fleming and Schmiegelow 2003). These species are not regularly seen on or near seismic lines, suggesting that there may be a critical width needed for disturbance-tolerant species to be attracted to linear features or well pads (Machtans 2006). However, as with forest specialists the critical width and level of vegetation recovery needed to create habitat for early successional species are not known for the boreal forest. Studies in deciduous forests of the United States indicate that lines wider than 30 meters are far more likely to have early successional species than 12- to 20-meter-wide lines (Anderson et al. 1977; Confer and Pascoe 2003).

Edge Effects

The small spatial scale of most energy sector disturbances creates a disproportionate amount of edge habitat per unit area disturbed. Edge is the transition zone between two habitat types, and various ecological conditions change near edges that may alter habitat suitability for songbirds (Murcia 1995), particularly in forests. Twenty-five studies have assessed the abundance of bird species near versus far from energy sector disturbances in forested habitats across North America (table 6.1).

Ideally, all studies would report their results in a common metric such as density, which would allow direct comparison of the effect size generated by energy sector edges of different types. We tallied the number of positive, negative, and neutral responses of edge for boreal forest passerines because inconsistencies in reporting preclude a meta-analysis. Four species were regularly more abundant in forest interiors than edges in at least two studies, and six species have a single study reporting a negative edge effect. Negative responses to edge were most commonly reported for ovenbird and black-throated green warbler. Six species have inconsistent responses to edge, sometimes being positive and other times negative (e.g., American redstart). Of these, most positive effects are observed at edges of pipelines, power lines, or gravel roads, and negative responses are most often reported near paved roads. Positive edge responses in at least two locations occurred for twelve species, with an additional twenty-two having a single study reporting a positive edge effect. Red-eyed vireo (*Vireo olivaceus*) is the species most consistently found at edges across North America.

TABLE 6.1. Forest passerines in boreal forest of western North America that show a negative response to forest edge created by roads, power lines, pipelines, seismic lines, or well pads. Values are the number of studies examining whether a neutral ($E = I$), positive ($E > I$), or negative ($I > E$) response to edge (E) versus interior (I) forest was reported. Studies that examined more than one type of disturbance could be counted twice.

Species	Scientific Name	$E = I$	$E > I$	$I > E$
Typically Negative Response to Edge				
Ovenbird	*Seiurus aurocapilla*	7	0	9
Black-throated green warbler	*Dendroica virens*	1	0	4
Red-breasted nuthatch	*Sitta canadensis*	1	0	2
Winter wren	*Troglodytes troglodytes*	3	0	2
Bay-breasted warbler	*Dendroica castanea*	2	0	1
Blackburnian warbler	*Dendroica fusca*	3	0	1
Connecticut warbler	*Oporonis agilis*	0	0	1
Golden-crowned kinglet	*Regulus satrapa*	2	0	1
Hermit thrush	*Catharus guttatus*	3	0	1
Yellow-bellied flycatcher	*Empidonax flaviventris*	0	0	1
Occasional Negative Response to Edge				
Black-capped chickadee	*Poecile atricapillus*	3	1	4
American redstart	*Setophaga ruticilla*	6	3	3
Black & white warbler	*Mniotilta varia*	8	1	3
Blue-headed vireo	*Vireo solitarius*	3	1	1
Mourning warbler	*Oporornis philadelphia*	0	1	1
Swainson's thrush	*Catharus ustulatus*	4	1	1

Beyond Linear Features

Polygonal disturbances, such as compressor stations, booster stations, and well pads, also create small patches of early successional habitat 0.5–2 hectares in size. We lack studies that identify the spatial scales at which avoidance of well pads results in a reduction in local population size. Preliminary findings suggest that many boreal passerines avoid using well pads as part of their territory (Lisa Mahon, University of Alberta, unpublished data). Only fifteen of fifty possible species occurred within point counts conducted at abandoned well pads in mixed-wood forest. Lincoln's sparrow,

Tennessee warbler (*Vermivora peregrina*), and chipping sparrow were the dominant species that would use well pads.

Patterns of habitat use by passerines in the boreal forest at energy sector disturbances are somewhat expected. Studies on bird response to forest harvesting reveal that the species using energy sector disturbances are similar to those on regenerating cut blocks less than 10 years of age, although direct comparisons have not been done (Song 2002). Forestry professionals have recognized that their operations alter habitat structure and composition in the short term but have worked to reduce the time it takes for habitats to recover. Over time, bird communities typically return to what is expected relative to natural disturbance, with notable exceptions of fire-dependent species (Hobson and Schieck 1999).

The value of energy sector sites for early successional bird species could be improved by putting more effort into retaining some habitat structure on disturbed sites (Yahner 2008). For example, Lee and Boutin (2006) demonstrated that only 8.2 percent of conventional seismic lines in boreal Alberta reached 50 percent cover of woody vegetation after 35 years. The lines that did recover were generally found in upland forests, whereas lines in lowland rarely recovered. Line recovery is limited because they are repeatedly used for access by industry and recreationalists. A substantial percentage (about 25 percent) of seismic lines never return to forest as they transition into roads, pipelines, or buildings. Well pads show a similar pattern. In northeast Alberta about 36 percent of 15-year-old well pads in upland forest have fewer than 1,000 tree stems per hectare, 24 percent have 1,000–6,000 stems per hectare, and 39 percent have more than 6,000 stems per hectare. A typical regenerating cut block is expected to have about 10,000 stems per hectare (MacFarlane 2003). Recovery of energy sector sites could follow a similar trajectory to clearcuts, but current legislation and industry practices prevent rapid recovery. The desire to minimize erosion effects has resulted in a policy of planting grass on well pads in Alberta. Seeding of grass to limit erosion, loss of natural propagules, removal of nutrients due to total removal of forest understory vegetation, and deliberate compaction of soil are all factors that limit regrowth on well pads. Research is being done in boreal Alberta to identify industry best practices to encourage tree growth, but broad implementation of these practices is lacking.

What Bird Abundance Tells Us

Bird abundance can provide a false signal about population growth if habitats in which birds settle are ecological traps that result in poor reproduction (Pulliam 1996; Bayne and Hobson 2002). The potential for reduced population growth caused by reduced vital rates around energy sector sites is a concern. Only one study has been conducted on nesting success of boreal birds in relation to energy development. Ball et al. (2009) studied nest success of forest passerines along pipelines in two areas of the boreal forest. Their first study was in a remote area in the Northwest Territories, where the pipeline is the only major source of human disturbance. They hypothesized that nesting near an edge might depress nest success if predators alter their behavior to forage near or move along pipeline right-of-ways. Over 2 years and about 700 nests they found no evidence of increased predation for birds nesting closer to the pipeline. A second area was examined where both energy development and timber harvest were prevalent. They hypothesized that nesting near an edge in a more fragmented landscape might increase risk of predation because landscape-level edge influences the type and frequency of predators. Again they found little evidence of an increase in predation rate between the edge and the interior. The dominant nest predator in both areas was the red squirrel (*Tamiasciurus hudsonicus*), whose abundance was similar close to and far from pipeline edges.

Avian and Human Perspectives on Energy Sector Activity in Remnant Grasslands

From a human perspective, a well pad or pipeline cut in a forest brings a stark visual change. But people may walk across disturbed sites in native prairie and scarcely notice the difference. Yet seemingly subtle changes in open systems may severely affect grassland birds because they select habitats based on fine-scale vegetation features (Robbins and Dale 1999; Green et al. 2002; Wheelwright and Rising 2008). Some biologists refer to native grasslands as old-growth prairie because the permanent and negative impacts of breaking the sod of native prairie are similar to or worse than harvesting old-growth forest.

Temperate grasslands have the highest ratio of converted to protected area of any major biome (Hoekstra et al. 2005). Estimated grassland losses since the mid-1800s exceed 75 percent in most areas of North America (Samson and Knopf 1994; Gauthier and Wiken 2003). Corresponding declines in grassland birds are the largest of any bird group (Askins et al. 2007; Brennan and Kuvlesky 2005; Sauer et al. 2008), and they occupy a disproportionate number of places on protected species lists. Many shrubsteppe bird species are also in sharp decline (Knick et al. 2003). Energy developments are becoming a dominant land use in what remains of these open habitats (chap. 2). The footprint associated with energy development at current densities can be 5–12 percent of the landscape, and linear features increase by a kilometer for each new well (Government of Canada 2008). Further loss and fragmentation associated with this new land use will only exacerbate ongoing conservation challenges.

Most open-country birds are migratory and can be divided on the basis of four characteristics: those that need or at least accept shrub cover and those that avoid it, those tolerant of nonnative vegetation and those that occupy only native prairie, those that need large tracts and those that reproduce successfully in small habitat fragments, and those endemic to the West versus those distributed across North America. A species with a restricted distribution, a need for shrubs, an aversion to nonnative plants, or an avoidance of edge is most likely to respond negatively to disturbances created by energy development. Open-country birds with a known negative response to energy share one or more of these characteristics (table 6.2).

More grassland birds have these vulnerable characteristics, but in only a few cases has this response been examined and confirmed because few studies in open habitats have evaluated whether birds forage, sing, or nest on disturbed energy sites. Instead, researchers typically record bird abundance near and far from development. Findings may document the avoidance of otherwise suitable habitat, but little has been learned from these studies about the mechanisms underlying the biological response.

Avoidance of Infrastructure

Studies on bird abundance near and far from energy infrastructure suggest that some grassland species avoid developments. In Minnesota, planted grasslands without wind turbines or maintenance trails, and areas located 180 meters from such development, supported higher densities (261–312 males per 100 hectares) of grassland birds than areas within 80 meters of turbines (58–128 males per 100 hectares; Leddy et al. 1999). In North Dakota, the grasshopper sparrow avoided wind turbines out to 200 meters, which resulted in a bird abundance 45 percent lower at experimental sites than at control sites after development (Shaffer and Johnson 2008).

Infrastructure associated with conventional oil and gas development contains fewer tall structures than wind energy facilities, but negative responses by many bird species persist. Chestnut-collared longspur, Sprague's pipit, and Baird's sparrow each showed a nonsignificant pattern of lower

TABLE 6.2. Summary of known or inferred negative (–), positive (+), no (=), or mixed (+/–) responses by open-country birds of western North America to factors associated with oil, gas, or coal extraction activities.

Species	Concern Score[a]	Restricted to Western North America[b]	Linear	Edge	Alien Plants	Shrub
Baird's sparrow (*Ammodramus bairdii*)	17	Yes	–[c,d]	–[e]	–[f]	–[f]
Sprague's pipit (*Anthus spragueii*)	16	Yes	–[c,d]	–[e]	–[g]	–[g]
Brewer's sparrow (*Spizella breweri*)	13	Yes	–[h]	–[j]	–[j]	+[k]
Sage sparrow (*Amphispiza belli*)	13	Yes	–[h]	–[j]	–[j]	+[l]
Chestnut-collared longspur (*Calcarius ornatus*)	13	Yes	–[c,d]	–[e]	–[m,n]	–[n]
Grasshopper sparrow (*Ammodramus savannarum*)	12	No	=[e]	–[o]+[q]	–[p]	–[p]
Vesper sparrow (*Pooecetes gramineus*)	11	No	+[c,d]	–[r]	=[q,r]	+/–[r]

[a]Continental Combined Score (Panjabi et al. 2005). Scores are based on range, population numbers, trend, and perceived threats. Scores range from 4 for a widespread, numerous, increasing species to 20 for a species of the very highest conservation concern. Birds known to respond negatively to some aspect of energy extraction are presented in order of decreasing score.

[b]Mengel (1970); [c]Linnen (2008); [d]Sutter et al. (2000); [e]Davis (2004); [f]summarized in Green et al. (2002); [g]summarized in Robbins and Dale (1999); [h]Ingelfinger and Anderson (2004); [i]Knick and Rotenberry (1995); [j]Knick and Rotenberry (1999); [k]Rotenberry et al. (1999); [l]Martin and Carlson (1998); [m]Lloyd and Martin (2005); [n]summarized in Hill and Gould (1997); [o]summarized in Ribic et al. (2009); [p]summarized in Vickery (1996); [q]Davis and Duncan (1999); [r]summarized in Jones and Cornely (2002).

abundance near minimal disturbance gas wells and associated trails in Saskatchewan compared with areas farther away (Linnen 2006). Reductions in these same three species were observed out to 50, 250, and 350 meters from traditional oil wells and their access routes in Alberta (Linnen 2008). Another Alberta study found that Sprague's pipit territories crossed trails less often than expected by chance (Hamilton 2009). Avoidance in landscapes crisscrossed by trails may greatly reduce available habitat.

Why grassland birds avoid roads and trails is unclear, but possible reasons include traffic, edge avoidance, and human presence. Grassland birds in eastern North America showed increased avoidance with increased traffic (Forman et al. 2002), but traffic levels on most energy access roads would be far lower. Studies in Wyoming sagebrush indicate that narrow low-use roads (10–700 vehicles per day) associated with gas wells had 39–60 percent fewer sagebrush obligate birds within 100 meters than at farther distances (Ingelfinger and Anderson 2004). However, narrow linear features do not need to be regularly traveled by vehicles to have an effect on birds. Nest densities and the abundance of some bird species were lower near narrow walking paths in Colorado grasslands (Miller et al. 1998). Whether birds perceive roads and trails as edge is hard to separate from other possible causes of avoidance, but Hamilton (2009) noted that Sprague's pipits whose territories were near trails often used the trail as a boundary. She postulated that narrow strips of bare ground or reduced vegetation would be clearly visible to an aerial singing bird. The grassland species that Linnen (2008) found avoided wells and associated trails (Sprague's pipit, Baird's sparrow, and chestnut-collared longspur) are all edge- and area-sensitive (Davis 2004).

Ashenhurst and Hannon (2008) also documented bird abundance on new (0.5–1.5 years old) and old (10–30 years) seismic lines and nearby controls in three open tundra habitats within the Kendall Island Migratory Bird Sanctuary, Northwest Territories. Abundance of all species except Lapland longspur (*Calcarius lapponicus*) was lower on old seismic lines than in reference sites in upland tundra habitats. Vegetative studies at the same site found that seismic lines had less lichen cover, more bare ground, and more vascular plant cover than reference sites, with changes persisting in the long term and modern lines causing just as much impact as old exploration. Many of these lines have not been traveled on by a vehicle in 30 years (Kemper and Macdonald 2009a, 2009b).

Chronic or episodic human visitation has been shown to cause many species of birds to alter song patterns, avoid otherwise suitable habitat, or interrupt incubation or mating (Baydack and Hein 1987; Gutzwiller et al.

1994, 1997; Miller et al. 1998; McGowan and Simons 2006). These studies were not specific to energy development.

As in forested habitats, few BACI studies have been conducted on open-country birds at a large enough spatial scale to document changes in populations. In an area of about 500 square kilometers within Canadian Forces Base Suffield, Alberta, a BACI study comparing the same areas at four well pads per 2.59 square kilometers and later with eight well pads per 2.59 square kilometers found a nonsignificant reduction in Sprague's pipit of 13 percent and a significant decrease of 21 percent for Baird's sparrow (Dale et al. 2009). Effects of energy-related disturbance are predicted to be at their worst when the species is under the greatest natural stress. A model for this study area (Brenda Dale, Canadian Wildlife Service, unpublished data) that includes drought effects on vegetation found that during normal precipitation, much of the area would be suitable for Sprague's pipit and that increasing well density from eight to sixteen per 2.59 square kilometers would result in only 5 percent of otherwise suitable habitat becoming unusable. However, during drought much less of the area would be suitable, and increasing the well density to sixteen per 2.59 square kilometers would result in 86 percent of the already limited suitable habitat becoming unusable, creating a potential limiting factor for the species.

Problems with Invasive Plants

Invasion of exotic plants may be an indirect and negative effect of energy development to grassland birds (chap. 7). Evidence suggests that development facilitates the introduction and subsequent spread of nonnative plants (Tyser and Worley 1992; Larson et al. 2001; Gelbard and Belnap 2003; Gelbard and Harrison 2003). Vehicles are a vector for seed dispersal and establishment along roadways (Von der Lippe and Kowarik 2007), and published reviews show that nonnative cover is higher at well pads, pipelines, and access routes than in native prairie (Bergquist et al. 2007; Rowland 2008). Nonnative vegetation matters to birds because it often differs in structure from native vegetation. Altered structure can have significant implications for birds by altering nest site availability and influencing the number of microsites where birds can locate and capture insect prey. For example, invasive brome grasses (*Bromus* spp.) die much earlier than native plants, and the abundant litter decomposes more slowly (Ogle et al. 2003). Crested wheatgrass (*Agropyron cristatum*) has less vegetation and less loose litter in the 10 centimeters nearest the ground, which results in greater amounts of bare ground compared with native prairie (Sutter and Brigham 1998; Christian and Wilson 1999). Within Alberta study blocks, cover of crested wheatgrass at the well pad and in adjacent prairie habitat increased with well density, but most Sprague's pipit territories contained no appreciable crested wheatgrass, and average values were lower than at reference locations (Hamilton 2009). This is consistent with the bird's limited use of the invasive plant (Robbins and Dale 1999).

Restoring Open Habitats

Energy developers contend that native prairie can be restored quickly and easily to its original condition after disturbance. Periodic natural disturbance helps maintain healthy grasslands and grassland bird communities, but energy-related disturbance is linear and long lasting. We know of no example of successful restoration to the original native plant community. Reclaimed coal mines in eastern North America have been well studied and support a number of grassland birds (Whitmore and Hall 1978), but reproductive success at these sites was too low to maintain positive population growth (Wray et al. 1982). Abundance of vesper sparrow was lower on reclaimed and unreclaimed mine spoil than on control sites in South Dakota and Wyoming (Schaid et al. 1983). Another revegetated surface coal mine in Wyoming exhibited marked differences in vegetation structure and associated bird communities 4 years after reclamation (Parmenter et al. 1985). Some sites may not even fully revegetate. There was more bare soil on recent prairie well pads and pipelines at Canadian Forces Base Suffield in Alberta than on reference sites, including most wells more than 30 years old (Rowland 2008). Even if areas disturbed by energy development are revegetated, bird communities will not fully recover, given that native grasslands support a higher frequency and abundance of endemic grassland birds than planted grasslands (Johnson and Schwartz 1993; McMaster and Davis 2001). Restoration remains an unproven hypothesis and should not be viewed as a panacea for conservation and management of grassland birds.

The Emerging Issue of Noise

Timber harvest, oil and gas drilling, and construction of energy infrastructure are intense, short-term events that birds may avoid for a period of time.

However, the infrastructure built by the energy sector has high maintenance needs, which results in more frequent human activity at sites than would occur in a regenerating forest cut block or native pasture. Repeated human use creates physical, visual, and acoustic disturbances that birds may avoid indefinitely. Of particular concern is anthropogenic noise because most songbirds use acoustic signaling to communicate. Males sing to defend territories and attract female mates. Females chip to communicate with males about nest attendance. Nestlings beg to get fed. Anything that interferes with the ability of birds to communicate has the potential to reduce habitat quality. Energy development creates anthropogenic noise via increased traffic, construction equipment to build and maintain infrastructure, and engines to compress and transport oil and gas through pipelines.

One of the most studied aspects of noise pollution is the effect of road noise on passerine birds. A road with 10,000 vehicles per day has an estimated maximum zone of disturbance of 125–190 meters where bird density is 30–100 percent lower than within habitat interiors (Reijnen et al. 1997). Effects were most pronounced in grassland habitats. Increasing traffic levels fivefold resulted in a threefold increase in area affected. The potential impact of such disturbances could be significant. For example, the major development route to energy reserves in boreal Alberta is Highway 63, where traffic levels range from 2,000 to 55,000 vehicles per day depending on location along the route (Alberta Transportation 2008).

Proximity and magnitude of noise are the two best predictors of bird response to this disturbance. An individual will return to an area, particularly if it is territorial, when the disturbance stops and noise dissipates. Of concern is when noise events coincide with breeding and a bird is forced to leave the nest, or nestlings, alone for extended periods. A potentially greater long-term risk is created by chronic noise disturbance. Sources of chronic noise in energy development include pump jacks, booster stations, wind turbines, and compressor stations that run continuously.

Birds may avoid otherwise suitable habitat if their communication is continually interrupted. Bird density was 1.5 times higher in boreal aspen forests with no anthropogenic noise than the same habitat beside a noise-generating compressor station (Bayne et al. 2008). One-third of all bird species detected in this study were less abundant within 300 meters of noisy areas. On the same study sites, Habib et al. (2007) demonstrated that male ovenbirds near noisy compressor stations were 15 percent less likely to attract a mate than males in quiet areas. Compressor stations generally produce noise levels between 75 and 90 decibels at the source, which is similar to a road with about 50,000 cars per day. More than 5,000 compressor stations exist in the province of Alberta alone and are only a small part of the noise-generating facilities linked to energy sector infrastructure.

Effects of sound in open habitats have not been studied explicitly but are potentially larger than those observed in forests. In Kansas, sound levels from compressor stations were 80–100 decibels at 100 meters and were clearly audible for 2 kilometers (Pitman et al. 2005). In Alberta, noise 50 meters from well sites generated by drilling (70 decibels) and maintenance activities (72 decibels) took 500 meters (EnCana 2007) to fall at or below the suggested maximum threshold of 49 decibels in the vicinity of breeding songbirds and raptors (Wyoming Fish and Game Department 2009) and was still greater than 25 decibels at the furthest distance measured (1.5 kilometers).

Just as they do with vegetation changes, generalist species may prosper from noise at the expense of intolerant specialist species of management concern. House finch (*Carpodacus mexicanus*) and juniper titmice (*Baeolophus ridgwayi*) increased and spotted towhee (*Pipilo maculatus*) decreased when noisy compressor stations were introduced into pinyon–juniper woodlands (LaGory et al. 2001). Francis et al. (2009) expanded this study and found the same species increasing, but added that mourning dove (*Zenaida macroura*), gray flycatcher (*Empidonax wrightii*), gray vireo (*Vireo vicinior*), black-throated gray warbler (*Dendroica nigrescens*), and spotted towhee nested farther from compressor stations than noiseless well pads. Surprisingly, nesting near a compressor station reduced predation risk, which is generally the largest source of mortality for passerine nests. Nests near compressor stations had higher survival (80.4 percent) than control sites (61.2 percent), which they attributed to predators avoiding areas of noise. Although it is a plausible explanation for this pattern, more of the nests monitored near compressors were of species tolerant of noise (i.e., house finch), which may have biased results if these species have nesting strategies that protect them from predation risk.

Conclusion

Energy extraction is a widespread and growing industry that broadly overlaps the boreal forest, prairie grasslands, and increasingly northern tundra of western North America. Yet we know little about its effects on songbirds, particularly at the population level, and whether these impacts are additive or synergistic relative to those caused by other human land uses. The few studies to date show strong evidence of direct mortality and avoidance of energy infrastructure, but also point at ways the energy sector can reduce impacts. However, best practices that focus on restoration or minimization of the size and duration of energy footprint are either not tested or poorly tested. For the energy sector to mitigate and reduce their impacts, ecologists need to continue to work to identify useful best practices. Wyoming Fish and Game Department (2009) recommendations are an example of best practices that will reduce impacts. Importantly, many of these recommendations were developed by leading energy sector companies and have been adopted by industry. However, whether best practices are enough to conserve all birds of concern is debatable. Targeting areas where conservation protection is the paramount goal is becoming increasingly important in many areas of North America where the energy sector is operating. In many cases, excluding development may be the only way to ensure that energy sector activities do not affect sensitive bird populations.

The experiments that are being done must be monitored more effectively. At a local scale, behavioral avoidance is often subtle. Therefore, methods of avian sampling must be accurate and precise to fully elucidate impacts. Traditional approaches of using imprecise point counts are insufficient to measure the magnitude of effects caused by energy development (e.g., Bayne et al. 2005a). Far more work is needed to identify whether vital rates of birds are affected by energy sector development and to test how population dynamics will be influenced by continuing development. In particular, the impacts caused by adult and juvenile mortality due to collisions must be put in the context of population growth. Current research has focused on examining one energy sector issue at a time. Although this approach can be effective at elucidating mechanisms behind bird changes in behavior, it will do little to answer the question of which species are at the greatest risk. To do this we need to ensure that behavioral processes are linked to population and community dynamics. We also need to combine the results from different studies into a coherent whole to measure the combined impact of all these different energy sector activities on songbirds. Alternatively or perhaps in concert, more effort must go into monitoring populations in a BACI framework at large spatial scales that integrate all the changes in ecosystems brought about by energy sector development.

We must establish priorities to facilitate effective research (e.g., chap. 3). Every management action will increase some species and decrease others. It is important that the losers can afford to lose (i.e., are widespread, abundant, and not declining due to other human activities). In grasslands all the species with documented effects (prairie chickens, Sprague's pipit, grasshopper and Baird's sparrows, chestnut-collared longspur) are declining species, and most are ranked as being of above-average conservation concern by Partners in Flight (table 6.2). From a conservation perspective, they cannot afford to lose. Although conservation targets (Rich et al. 2005) provide goals of what is desired for birds, the ability of ecologists to provide concrete recommendations of how to achieve these desired states remains inadequate. To achieve this goal, ecologists need to report their findings in a manner that can be better quantified so that meta-analyses and synthesis of larger datasets can be used with future simulation tools to make better decisions about whether or where particular developments can occur while maintaining bird populations.

While continuing to document and understand effects and mitigation tools, we also need to determine how large an impact the energy sector is going to have on bird populations over time. We need to move beyond documenting effects to estimating population impacts. Cumulative impact research is needed to yield landscape-level estimates for broad-scale conservation planning. We need to recognize that energy extraction is not the only land use. Stakeholders remain ill-equipped to maintain bird populations yet still allow the use of resources needed by society until the cumulative impacts of energy, forestry, agriculture, and other activities are quantified in a coordinated manner.

Chapter 7

Invasive Plants and Their Response to Energy Development

PAUL H. EVANGELISTA, ALYCIA W. CRALL,
AND ERIN BERGQUIST

Increases in human developments, travel, and trade greatly facilitate the movement of organisms to locations where they do not naturally occur (Mack and Lonsdale 2001). These organisms are called nonnative species, and *exotic*, *alien*, *introduced*, and *nonindigenous* are common synonyms for this term. Once a nonnative species is introduced to a new area, it may no longer be susceptible to the population controls that coevolved with it in its native habitat. In some cases, nonnative species may have a benign relationship with native species, resulting in little effect on ecosystem processes. In other cases, the nonnative species may have a competitive advantage over native species and spread quickly, becoming invasive. Excluding cultivars, 10–20 percent of the estimated 50,000 nonnative species introduced to the United States (Pimentel et al. 2005) will become invasive (Chornesky and Randall 2003). Invasive species affect native species by direct competition and by indirectly altering ecosystem processes such as food webs, hydrology, nutrient and decomposition cycles, and natural disturbance regimes (Vitousek et al. 1997a, 1997b; Mack et al. 2000; Mooney et al. 2005). Most invasions detrimentally affect the environment, human health, or the economy, and in some cases native species are extirpated or become extinct.

Management of invasive species has become a high priority for scientists and resource specialists in western North America and worldwide. Second to habitat loss, invasive species pose the greatest threat to global biodiversity. Invasive species adversely affect more than 40 percent of species listed under the U.S. Endangered Species Act (Randall 1996; Wilcove et al. 1998) and threaten up to 80 percent of imperiled species worldwide (Armstrong 1995). By modifying ecosystem processes, invasive species magnify their effect far beyond the level of individual species, resulting in global impacts. Collectively, impacts result in overwhelming losses to national economies. Invasive species cost U.S. taxpayers $120 billion annually in lost revenues, maintenance, and control efforts (Pimentel et al. 2005); other reviews suggest that economic costs are much greater and too difficult to quantify (Lodge et al. 2006). Most ecologists and resource managers agree that long-term environmental effects and economic costs are yet to be realized, and that some impacts are simply immeasurable.

This chapter cautions readers about the potential for energy development to imperil native species and degrade ecosystems by facilitating the proliferation of invasive species. Objectives of this chapter are to educate readers on ways in which energy development acts as a conduit for invasive species, inform decision makers about the threat invasive species pose to native and intact western ecosystems, direct policymakers to forgo development in landscapes where other natural resource values outweigh those of extraction, and encourage resource managers to limit impacts within developments by requiring interim reclamation and restorative activities. We provide an overview of patterns in plant invasions, synthesize resulting impacts on ecosystem processes, present findings from a case study in Wyoming, and discuss what can be done to reduce future impacts on species and ecosystems. We present a case study rather than a synthesis of the literature because surprisingly little research has been published on the effects of energy development on the distribution and spread of invasive species. The critical experiments necessary to quantify the scales at which energy developments facilitate invasions simply have not been done (chap. 3).

Patterns of Plant Invasions

Most nonnative plant introductions are intentional and include species brought to the United States for agriculture, horticulture, and environmental reclamation. In the United States, an estimated 5,000 nonnative plant species have escaped and persist in natural landscapes (Morse et al. 1995), a large number considering that only 17,000 species are native (Morin 1995). Once introduced, most nonnative species exhibit lag times characterized as initial periods of inactivity followed by a sudden expansion in its population or range (Crooks 2005). Lag times vary by species, with some lasting only a few years and others for decades. Melaleuca (*Melaleuca quinquenervia*) was introduced to Florida in late 1800s, but its population was not considered invasive until the 1950s (Ewel 1986). Lag times provide nonnative species with the opportunity to establish adequate reproducing populations (D'Antonio et al. 2001), improve genetic composition (Lee 2002), or allow environmental conditions to become favorable (Rejmanek et al. 2005). Once an invasion begins, a species can advance at alarming rates. Patterns of plant invasions are difficult to generalize across taxa, but an invasion is typically facilitated by species traits, ecosystem vulnerability, and anthropogenic influence (Thuiller et al. 2006). Here we review the biology of these three attributes that drive new invasions.

Species Traits

Most invasive species exhibit unique physiological traits that provide them with a competitive advantage over their native counterparts. These traits may enhance an invader's ability to exploit resources (Stohlgren et al. 2003), survive under adverse conditions (Blackburn et al. 1982), successfully reproduce (Horton 1977), or modify ecological conditions to their advantage (D'Antonio and Hobbie 2005). Successful invaders usually exhibit multiple traits that collectively enable them to disperse across a wide range of ecological conditions and environmental gradients (Lodge 1993; Pysek and Richardson 2007). Tamarisk (*Tamarix* spp.), a native of Eurasia, the Mediterranean, and northern Africa, provides a telling example of how a species' traits can contribute to its success as an invader. First introduced in the early 1800s, tamarisk is known for its remarkable ability to spread, adversely affecting native flora, wildlife habitat, and hydrologic processes throughout the southwestern United States (Christensen 1962; Harris 1966). A prolific seed producer, tamarisk generates up to 600,000 seeds annually that are dispersed over long distances by wind or water (Robinson 1965). Reproductive opportunities are further enhanced by its ability to develop adventitious roots from branches and clippings (Horton 1977). Members of the genus are facultative phreatophytes able to extend their root systems as deep as 50 meters, desiccating floodplains and water tables (Blackburn et al. 1982; Pinay et al. 1992). Tamarisk uses water more efficiently than native cottonwoods (*Populus* spp.) and willows (*Salix* spp.) (Busch and Smith 1995; Cleverly et al. 1997) and, in some cases, can lower water tables to depths that are out of reach to native plants. Tamarisk tolerates a wide range of conditions, including drought (Blackburn et al. 1982), flooding (Irvine and West 1979), and stem removal (Horton 1977). Infestations can alter ecosystem processes and modify habitat conditions to tamarisk's advantage. Dense canopies can shade sunlight, and its leaf litter accumulates in thick mats in the understory, inhibiting the germination and growth of native plants. Tamarisk has a high tolerance to saline soil. Salts are assimilated by its roots and deposited in the leaf litter, where they accumulate in concentrations too high for many native species (Carman and Brotherson 1982; DiTomaso 1998).

Ecosystem Vulnerability

Vulnerability of an ecosystem to potential invasion is often correlated with biotic and abiotic conditions. Conditions include the diversity of the invaded ecosystem, its resource availability, and the occurrence of natural and human disturbances. Ecosystems with high native plant diversity are typically less vulnerable to invasion, implying that intact native plant communities occupy a greater proportion of ecological niches while securing available resources (Elton 1958; Tilman 1999). Studies have repeatedly demonstrated that natural and human-caused disruptions of native flora result in a higher incidence of invasion (Evangelista et al. 2004). New research also suggests that intact and stable ecosystems are not as impervious to invasion as once thought. In some cases, ecosystems with high native plant diversity and abundant resources are being invaded at rates comparable to those that have been disturbed (Levine and D'Antonio 1999; McKinney 2002; Stohlgren et al. 2003). Findings highlight the ability of some invasive species to outcompete native species.

Resource availability is a primary determinant of the establishment and persistence of native and nonnative plants. Some species need precise environmental conditions to survive, whereas others can endure greater environmental variability. Interactions between a potential invader and resource availability in different habitats provide clues as to how a species will respond. Generalist species, capable of exploiting resources under a variety of conditions, may invade multiple habitats, resulting in a wide range of dispersion (Evangelista et al. 2008). Japanese honeysuckle (*Lonicera japonica*) can readily establish itself in open areas with high sunlight or in the shaded understory of mature forest (Sather 1992). Other invaders are specialist species, limited to specific habitats and conditions. Closed canopies limit some specialists to habitat edges with high light intensity (Brothers and Spingarn 1992).

Water availability in the West is a limiting resource that provides an opportunity for the establishment of invasive species that use water resources more efficiently than native species (Blackburn et al. 1982; Levine et al. 2003). A native to Eurasia, yellow starthistle (*Centaurea solstitialis*), was introduced from Chile in the mid-1800s as a contaminant of alfalfa. Found in at least twenty-three U.S. states (Maddox et al. 1985), yellow starthistle grows as a long-lived winter annual that germinates during fall in dry uplands, pastures, and rangelands. Throughout winter and early spring, it develops a deep root system that provides a competitive advantage over native annuals that germinate in spring. Once established, starthistle can rapidly deplete soil moisture from the ground surface, to the detriment of native seedlings. Its root system allows the plant to continually exploit water resources throughout the year, using up to half of the annually stored soil moisture (DiTomaso et al. 2003).

Anthropogenic Influence

Human disturbance is the most influential mechanism of plant invasion, providing numerous opportunities for invasion each day in the United States (Lodge et al. 2006). Pathways for new invasions include activities related to commerce (Kiritani and Yamamura 2003), transportation (Gelbard and Belnap 2003), and recreation (Rahel 2002). Road construction for energy development and other operations is a dominant feature of human-altered landscapes that facilitates invasions through multiple mechanisms. Clearing of vegetation and topsoil during road construction and maintenance opens ecological niches and improves accessibility to water, sunlight, and soil nutrients (Trombulak and Frissell 2000). Vehicles transport seeds and disperse them along roadsides, where they can germinate and establish new populations with ease (Schmidt 1989; Lonsdale and Lane 1994). Once established, invasive species typically are more resistant to roadside conditions than native species. Mowing, herbicide treatments, and soil compaction often prevent the reestablishment of natives while having only minimal effects on invaders (Forman and Alexander 1998; Gelbard and Belnap 2003). Once a seed source is established, an invasive species can infest almost any disturbed site.

Availability of soil nutrients, particularly nitrogen and phosphorus, is a primary determinant for many plant invasions (Bashkin et al. 2003; Ehrenfeld 2003). Disturbance to the soil surface can disrupt nutrient cycling and availability, and potential invaders may be quick to respond to the

disruption and to capitalize on nutrient surpluses. Such events are impli-
cated as one of the primary causes of successful invasions (Lozon and
MacIsaac 1997; D'Antonio et al. 1999). Increases in nutrient resources typically favor fast-growing annual species with a persistent source of seeds
that allows them to rapidly dominate disturbed sites (Huenneke et al.
1990; Baskin and Baskin 1998).

Impacts on Ecosystem Processes

Direct and indirect effects of invasive species on native species and ecosystem processes are wide ranging and often amplified over space and time
(chap. 3). Invasive species can cause native species to go extinct, which in
turn has direct consequences for other species that depend on them. Chestnut blight (*Cryphonectria parasitica*, an Asian fungus) was brought to the
United States on trees imported from Japan in the early 1900s. By 1950,
the American chestnut tree (*Castanea dentata*), along with seven moth species dependent on chestnut for their survival (Opler 1978), had disap-
peared from 3.5 million hectares of forested land (Roane et al. 1986). Invasive species also can reduce native biodiversity indirectly through changes
in ecosystem processes. The extent to which an invasive species alters ecosystem function depends on similarities in traits between the invader and
resident native species (D'Antonio and Hobbie 2005). Such was the case
with fayatree (*Myrica faya*), an invasive species brought to Hawaii by Por-
tuguese immigrants that has since invaded volcanic sites with limited nitrogen availability. As a nitrogen fixer, fayatree increased available nitrogen
fourfold, drastically altering ecosystem processes and allowing other native
and nonnative plants to establish (Vitousek et al. 1987; Vitousek 1990).

Relationships between native species and their habitats can be permanently disrupted if an invasive species alters natural disturbance regimes.
Disturbances affect primarily resource availability, particularly light and soil
nutrients, and often provide ideal conditions for invasive species (Evangelista et al. 2004). Once established, some invaders can facilitate disturbances
to their benefit, as invasive grasses have altered fire regimes in the West and
throughout the world (D'Antonio and Vitousek 1992; D'Antonio 2000).
Invasive grasses produce more aboveground litter than native species, increasing the probability, extent, and severity of fires (Parsons 1972; Mack
1986). Unlike native grasses, many of these invaders are adapted to fire and
rapidly seed or sprout after the disturbance. Fires become even more frequent as invading annuals increase in dominance. This positive feedback
loop results in a monoculture of few species, instead of the heterogeneous

landscape representative of the natural system (Rosentreter 1994). One such invasive, cheatgrass (*Bromus tectorum*), becomes dry and highly flammable at senescence, resulting in increased fuel loads (West 1983). Sagebrush grasslands invaded by cheatgrass burn every 3–5 years (Whisenant 1990), compared with a historic fire return interval of 40–350 years (Baker 2011). An artificially high fire frequency inhibits the ability of native shrubs and perennial grasses to recover from disturbance (Belnap et al. 2000).

Case Study: Energy Development in Sagebrush Grasslands

Land use in the Powder River Basin (PRB) in eastern Wyoming and Montana is primarily cattle ranching with limited dryland and irrigated tillage agriculture. Native vegetation is sagebrush steppe and mixed-grass prairie interspersed with occasional stands of conifers. Sagebrush steppe is dominated by Wyoming big sagebrush (*Artemisia tridentata wyomingensis*) with an understory of grasses and forbs. Plains silver sagebrush (*A. cana cana*) and black greasewood (*Sarcobatus vermiculatus*) co-occur with Wyoming big sagebrush in drainage bottoms. Shallow coal seams that underlie the PRB provide an abundant source of coal bed natural gas (CBNG). An estimated 1.1 trillion cubic meters of natural gas is recoverable in the PRB, and drilling has increased exponentially since the late 1990s, exceeding 35,000 wells. The CBNG is extracted by drilling wells, removing groundwater, and transporting gas through a network of buried pipelines. Well pads containing two to five well heads are situated on 1.6-hectare clearings that are typically connected by access roads, gas pipelines, power lines, generators, and a mechanism for water disposal. The groundwater that is pumped out is called coproduced water and varies widely in quantity and quality (Rice et al. 2000). One well may discharge 4,000–76,000 liters per day, which is commonly stored in containment ponds and surface reservoirs. The case study presented here focuses on invasive plants, but the coproduced water (Zou et al. 2006), which provides larval habitat for mosquitoes that vector disease, has been linked to the spread of West Nile virus (Flaviviridae), an invasive species that affects the growth of native sage-grouse (*Centrocercus urophasianus*) populations (Walker and Naugle 2011).

Methods

The only comprehensive study of invasive plants and energy development published to date used a control–impact design (Bergquist et al. 2007). This study compared four types of disturbance associated with CBNG to control sites without development to evaluate patterns in the distribution and richness (i.e., number of species) of nonnative plants. Four types of disturbances evaluated were surface disturbances including roads, pipelines, and dams; well pads within cleared areas containing individual well heads and associated infrastructure; discharge sites that were primary streams where coproduced water is released; and areas of secondary disturbance that were less than 50 meters from development activities. Non-CBNG control sites were located more than 50 meters from development activities.

Sampling was conducted within thirty-six modified Forest Inventory and Analysis plots, a multiscale circular sampling method commonly employed by the U.S. Forest Service for annual evaluations of the condition, change, and trends in ecosystem dynamics (Frayer and Furnival 1999). Modified Forest Inventory and Analysis plots were set up according to Bull et al. (1998), with each plot including four circular subplots. Each subplot had three 1-square-meter quadrats located along designated transect lines. Subplots were further stratified between those that had ($N = 64$) and had not ($N = 80$) been sprayed with herbicide in the previous year to control for invasive species. Using a 2.5-centimeter-diameter corer, soils were sampled to a depth of 15 centimeters at the center of each subplot. Soils were analyzed for nutrient and chemical composition. Complete descriptions of methods and statistical analyses are available in Bergquist et al. (2007).

Findings

Development of CBNG in the PRB is facilitating the introduction of several new invasive plants and enhancing the establishment of a few others. Richness and overall coverage of invasive species generally increased and native species decreased with disturbance (table 7.1).

Patterns were accentuated when control sites and secondary disturbance sites were grouped and compared with other disturbance types. Discharge sites, which were affected primarily by changes in water availability and soil chemistry, showed only a minimal decrease in overall species richness but a significant increase in the proportion of invasive species.

Soil chemistry and nutrient content varied between disturbance types. Some averages changed very little, whereas others showed dramatic increases. As expected, trends with some measures follow the intensity of disturbance (table 7.2).

TABLE 7.1. The percentage cover (±SE) and richness (±SE) of native and invasive species in 144 subplots for each disturbance type (as modified from Bergquist et al. 2007).

			Disturbance Type		
	Control	Secondary Disturbance	Surface Disturbance	Pads	Discharge
Number of subplots	73	42	14	8	7
Total species richness**	22.5a (1.0)	24.0a (1.3)	20.1ab (1.8)	11.6b (3.5)	22.0ab (2.8)
Native species richness**	16.6a (0.8)	17.3a (1.0)	14.1ab (1.5)	7.8b (2.5)	12.6ab (1.7)
Invasive species richness	4.4a (0.2)*	5.1a (0.4)	5.2a (0.5)	3.4a (1.2)	8.0b (1.0)
Proportion of invasive species richness**	21.4a (1.1)*	22.2ab (1.6)	26.6abc (2.1)	32.5bc (7.6)	37.5c (3.7)
Total percentage cover**	56.1a (2.4)	60.9a (3.7)	48.7ab (5.9)	32.6b (12.3)	55.4ab (9.2)
Native species percentage cover**	38.8a (2.2)	39.5a (2.7)	28.1ab (5.1)	17.7b (7.5)	23.1ab (6.1)
Invasive species percentage cover	16.1a (2.0)	20.5a (3.6)	20.3a (4.1)	14.7a (8.9)	31.0a (8.4)
Proportion of invasive percentage cover	26.9ab (2.6)	27.9ab (3.8)	43.7a (7.7)	22.1b (12.6)	55.4a (10.9)

*Denotes significant difference from results of t tests comparing controls and combined disturbances.
**Denotes significant difference from results of t tests comparing the grouping of controls and secondary disturbances and the combined primary disturbances.
a,b,cDenotes significant differences in mean values between disturbance types (Tukey's test).

TABLE 7.2. Mean soil chemistry values ($\pm SE$) in 144 subplots for each disturbance type (as modified from Bergquist et al. 2007).

	Disturbance Type				
	Control	Secondary Disturbance	Surface Disturbance	Pads	Discharge
Number of subplots	73	42	14	8	7
Soil salinity (μS/cm)**	34.0[a] (7.6)*	87.2[a] (27.5)	70.7[a] (30.8)	181.6[a] (99.3)	125.6[a] (87.0)
Percentage soil carbon	1.4[ab] (1.4)	1.3[a] (0.1)	1.6[ab] (0.2)	1.0[a] (0.1)	2.2[b] (0.7)
Percentage soil nitrogen	0.11[ab] (0.1)	0.10[ab] (<0.01)	0.14[a] (<0.01)	0.06[b] (<0.01)	0.14[a] (<0.01)
1:1 pH	7.6[a] (0.1)	7.6[a] (0.1)	7.4[a] (0.2)	7.8[a] (0.3)	7.9[a] (0.1)
Ca (mg/kg)	2,718.0[a] (105.0)	2,942.0[a] (162.0)	2,730.0[a] (253.0)	2,536.0[a] (516.0)	3,398.0[a] (161.0)
Mg (mg/kg)	481.0[a] (21.0)	392.0[b] (23.0)	490.0[ab] (52.0)	672.0[a] (142.0)	562.0[ab] (87.0)
Na (mg/kg)	57.0[a] (7.0)	52.0[a] (4.0)	65.0[a] (9.0)	169.0[b] (50.0)	265.0[b] (100.0)
K (mg/kg)	249.0[a] (16.0)	165.0[b] (9.0)	194.0[ab] (17.0)	176.0[ab] (28.0)	222.0[ab] (49.0)
P (mg/kg)	6.5[ab] (0.4)	5.4[a] (0.5)	7.6[ab] (1.1)	6.3[a] (2.5)	11.3[b] (3.6)

*Denotes significant difference from results of t tests comparing controls and combined disturbances.

**Denotes significant difference from results of t tests comparing the grouping of controls and secondary disturbances and the combined primary disturbances.

[a,b,c]Denotes significant differences in mean values between disturbance types (Tukey's test).

Soil salinity, in particular, increased with disturbance. Well pads and discharge sites, two distinctly different types of disturbance, had the highest levels of soil salinity. Soil salinity at well pads was approximately five times greater than at the control sites and nearly three times higher at discharge than control sites. Discharge sites also had higher levels of soil nitrogen, carbon, phosphorus, and sodium, and well pads had lower levels of soil nitrogen and potassium than control sites and other disturbance types.

Contrary to expectations, subplots treated with herbicides had less overall coverage of native species and a greater proportion of invasive species than untreated sites (table 7.3).

Within control subplots, those treated with herbicides had less native species cover than the untreated subplots. Within disturbance subplots, those treated with herbicides had less native species cover and a greater proportion of invasive species than untreated subplots.

Species accumulation curves suggest that trends at the subplot level may also be occurring at a broader scale (fig. 7.1).

TABLE 7.3. Percentage cover ($\pm SE$) and richness ($\pm SE$) of native and invasive species in 144 subplots in sprayed and unsprayed areas for each disturbance type (as modified from Bergquist et al. 2007).

	Control		Disturbance	
	Sprayed	Unsprayed	Sprayed	Unsprayed
Number of subplots	35	38	29	42
Total species richness	22.1 (1.3)	22.9 (1.5)	23.1 (1.5)	20.6 (1.4)
Native species richness	16.3 (1.1)	16.9 (1.3)	16.5 (1.2)	14.2 (1.1)
Invasive species richness	4.2 (0.2)	4.5 (0.3)	5.4 (0.5)	5.1 (0.4)
Proportion of invasive species richness	4.2 (0.3)	4.5 (0.4)	5.4 (0.5)	5.1 (0.4)
Total percentage cover	49.8 (3.3)*	62.0 (3.3)	54.2 (5.8)	55.2 (3.4)
Native species percentage cover	31.3 (2.7)*	45.6 (3.0)	26.5 (3.5)*	37.8 (2.9)
Invasive species percentage cover	16.7 (2.8)	15.5 (3.0)	27.2 (4.4)*	16.5 (3.0)
Proportion of invasive percentage cover	31.1 (3.9)	23.1 (3.5)	43.2 (5.4)*	26.1 (4.1)

*Denotes significant difference from results of *t* tests comparing sprayed and unsprayed areas in each disturbance type.

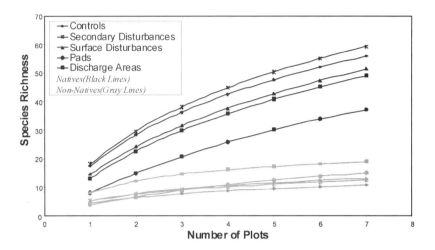

FIGURE 7.1. Species accumulation curves comparing richness of native (black lines) and invasive (gray lines) plants in four types of disturbance associated with energy development and in control sites without development (modified from Bergquist et al. 2007).

Curves show greater native diversity in control and secondary disturbances compared with the other three disturbance types. Shapes of curves also indicate the level of diversity and distribution across the landscape, with flattening curves indicating low diversity and commonness of nonnative species. Rising curves show high diversity of native species and their rarity across the landscape. Sampling identified 24 invasive and 147 native species. The three most common invasive species were cheatgrass, Japanese brome (*Bromus japonicas*), and desert madwort (*Alyssum desertorum*). Findings show that energy development is facilitating invasions, but disturbances associated with CBNG and their effects on ecosystem processes must be considered cumulatively. The mere presence of the three most common invasive species in 45–60 percent regardless of sampling location suggests that activities other than CBNG are also enabling new species to invade the PRB.

Implications for Managing Energy Development

Findings suggest that CBNG development facilitates the establishment and subsequent invasion of nonnative plants into sagebrush grasslands (Bergquist et al. 2007). Invasions resulting from this type of energy development

Photo 1.1. Photo credit: Mark Hebblewhite

The mere thought of the West conjures up mental images of wide-open spaces that support world-class wildlife populations.

In this decade, myriad studies have documented the cumulative impacts from energy development to sage-grouse, mule deer, caribou, and other western icons.

Photo 3.1

Photo 3.2. Photo credit: Lalenia Neufeld/Parks Canada

Photo 4.1

The scale of conservation must be analogous to that of development to maintain the large and intact sagebrush, forest, and grassland landscapes on which wildlife populations depend.

Photo 5.1. Photo credit: Brett L. Walker

Photo 5.2. Photo credit: Kurt Forman

Photo 6.1. Mark Hebblewhite

But in the midst of increasing energy demand, open space is at a premium, and poorly placed energy developments threaten our wildlife heritage.

Photo 7.1. Photo credit: Mark Gocke, WGFD

Photo 8.1. Photo credit: Hall Sawyer

Direct impacts of habitat loss from an individual well pad are small compared with indirect impacts that accumulate across the landscape. The human footprint of energy development in sagebrush habitats has resulted in the loss of critical mule deer winter range.

Photo 9.1. Photo credit: markgocke.com

Photo 10.1. Photo credit: Mark Hebblewhite

Energy development in the boreal forest has had similar and devastating effects on wildlife. Forest clearing, seismic lines, and energy roads fragment forests and provide conduits that facilitate increased predation by wolves on imperiled caribou populations.

Photo 11.1. Photo credit: John E. Marriott

Photo 11.2. Photo credit: Mark Bradley

Photo 12.1. Photo credit: Fred Greenslade

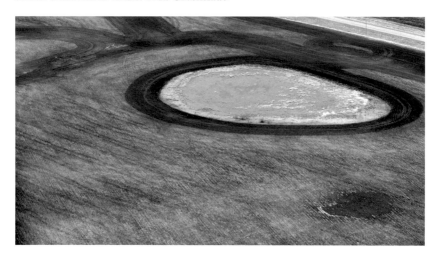

Tillage agriculture for biofuel production affects grassland-nesting birds that depend on these habitats for food and cover. Negative effects also include increased risk of wetland drainage after tillage. Grassland loss reduces waterfowl populations because the abundance of grass cover is directly related to duck nest success.

Photo 13.1. Photo credit: Fred Greenslade

Photo 13.2. Photo credit: Fred Greenslade

Photo 14.1. Photo credit: Barbara Cozzens

Renewable energy reduces our carbon footprint, but human disturbance from poorly placed wind and solar developments has unintended consequences for birds, bats, and other wildlife. The key to facilitating responsible development is in landscape planning and proper siting of all forms of infrastructure.

Photo 15.1. Photo credit: David Dodge, The Pembina Institute
(www.pembina.org)

Additional up-and-coming stressors include increases in uranium and coal mining
and steam-assisted gravity drainage exploration.

Photo 16.1. Photo credit: www.annesherwood.com

As tradeoffs between energy development and conservation unfold, one thing is clear: Additional impacts are inevitable. The solution for reducing impacts is in planning and implementing conservation in large and intact landscapes. At the heart of this solution are our rural ways of life. It's the people of the West—not governments—that will ultimately champion conservation.

Photo 16.2. Photo credit: Jeff Van Tine

are not surprising because findings are consistent with impacts resulting from other forms of human disturbance. Increasing demand for domestic energy dictates that extractive operations will continue to expand on public lands in the western United States and Canada. If the trends reported in Bergquist et al. (2007) are reoccurring, then managers should expect accelerated rates of invasions that result in severe impacts on ecosystems.

For resource managers, invasions pose a major challenge to conservation, with long-term consequences for many critical ecosystem processes. Resource managers need to consider the risks associated with invasive species when deciding where to permit developments on public lands. Our general scientific understanding of invasive species, when coupled with findings by Bergquist et al. (2007), presents a compelling and cautionary case for policymakers to forgo energy development in landscapes where other natural resource values outweigh those of extraction. Invasions at thousands of well pads and water discharge sites and along vast networks of roads and pipelines result in cumulative effects that degrade whole landscapes, especially when efforts to control invasive species appear ineffective (e.g., herbicides; Bergquist et al. 2007). Forgoing development in landscapes prioritized for conservation (chap. 12) is the only known way to alleviate the potential for new invasions (Mack et al. 2000).

The increasing extent of energy development (chap. 2) and cumulative footprint that accompanies it foreshadow the escalating ecological and economic costs of invasions. Cumulative effects severely limit the potential for future conservation because once an invasion begins, the resulting impacts are almost impossible to contain (Westbrooks 2004). The monetary costs of restoring public lands will fall on taxpayers because bonding requirements of industry are woefully inadequate. Alternatively, we may just have to live with the consequences of invasions, because examples of successful restoration on the scale of whole landscapes are limited for most ecotypes in the West, including sagebrush grasslands (Pyke 2011).

New introductions of nonnative species and subsequent invasions are expected in areas of energy development. Still, we encourage resource managers to require interim reclamation in landscapes where energy resources are developed, in hopes that appropriate steps will increase the success of future restoration. Reducing road densities and vehicle traffic on existing roads would greatly reduce major conduits of invasion. Washing vehicles to reduce the spread of nonnative seeds along roads, and using native seed mixes for reclamation, also may facilitate future restoration. Resource managers should be vigilant in identifying new introductions and reducing impacts while populations are small. Steps to achieve early detection and a plan for rapidly responding to new invasions should be included in management plans (Westbrooks 2004).

Once an invasive species is detected, its population should be quickly controlled to reduce impacts. Resource managers have chemical, mechanical, and biological options for containment (Westbrooks 2004). Each option has its own advantages and limitations, and careful consideration must be given to achieve the best management results (D'Antonio and Meyerson 2002). Chemical controls are commonly used but can be problematic because of their impacts on nontarget organisms, including humans (Mack et al. 2000). Herbicides can alter soil chemistry and have large-scale detrimental impacts if used in close proximity to water and may be toxic to animals. For new invasions, spot application using handheld sprayers can reduce harmful effects. Larger infestations are treated by vehicles or helicopters equipped with precision sprayers.

Mechanical control, including physical removal or a change in habitat conditions, is often used for extensive invasions but can be expensive and labor-intensive. Mechanical control may include hand removal, cutting, excavation, or burning. Large-scale operations usually disrupt the entire treatment area, and additional interim reclamation (e.g., seeding for erosion control) is almost always needed. Biological control methods rely on predators or parasites to reduce populations of invasive species. Control agents are typically a natural enemy from the native range of the invader. Such measures are usually a last resort and should be considered on a case-by-case basis, because some control agents do not survive the introduction and others attack nontarget organisms, creating additional problems (Simberloff and Stiling 1996).

Conclusion

The answer to additional development in the West is not "no" but rather "where" to extract resources as we strive to meet increasing domestic demand for energy. The sobering truth is that additional impacts on species and ecosystems are unavoidable. The future and primary role of science is to develop decision support tools with managers to identify where to site developments to protect other resource values that outweigh those of extraction, how to reduce impacts in landscapes where extraction takes place, and what investments in restoration yield the greatest biological benefits. Evidence to date suggests that development facilitates the spread of invasive species; however, the findings have limited utility to managers because the extent to which invasions reduce ecosystem function and imperil species remains unknown. Although theoretically justified, relying on the precautionary principle to restrict development everywhere is a futile exercise in energy policy.

The science of invasive species and energy development is in its infancy and parallels wildlife studies from just a few years ago (chap. 3). Studies of large ungulates have matured from documenting the avoidance of an individual well pad by elk (*Cervus elaphus*; van Dyke and Kleine 1996) to a mechanistic understanding of population-level declines in woodland caribou (*Rangifer tarandus tarandus*) as a result of increased predation by wolves (*Canis lupus*) along seismic lines cut through the forest (Sorensen et al. 2008; chap. 5). The challenge for plant ecologists will be to scale up from small-scale and short-term observational studies to long-term, before–after control–impact designs that help managers decide where to develop and still maintain functioning ecosystems and the species that depend on them. Scaling up is a difficult but necessary step if science is to provide managers with the ability to predict reliably the outcomes of alternative management scenarios. Large-scale spatial modeling will play an increasing role in the development of science support tools for management (Evangelista et al. 2004). Modeling is useful in assessing risk and potential distribution of invasive plants. By determining which ecosystems are most susceptible to plant invasions, land managers can prioritize areas for conservation, limit impacts in developed areas, and identify landscapes for restoration.

Chapter 8

Wind Power and Biofuels: A Green Dilemma for Wildlife Conservation

GREGORY D. JOHNSON AND SCOTT E. STEPHENS

Renewable or green energy is defined as energy generated from natural processes that are replenished over time; it includes electricity and heat generated from solar, wind, hydropower, biomass, geothermal resources, and biofuels derived from renewable resources. In 2006, around 18 percent of global energy use was derived from renewable sources (REN21 2008). The Obama Administration has made the development of renewable energy a top priority for economic expansion, to reduce U.S. dependence on fossil fuels and to lower greenhouse gas emissions contributing to climate change. More than twenty states have enacted laws requiring that a portion of the electricity supply come from renewable energy (American Wind Energy Association 2006). The U.S. Department of Energy (2008a) reports that it is technically feasible to generate 20 percent of the nation's electricity by 2030 from wind energy, and there is a goal of replacing 30 percent of transportation fuel consumption with renewable fuels by the year 2030. Although developing renewable energy sources is generally considered environmentally friendly, impacts on wildlife and their habitats can be associated with many forms of renewable energy (McDonald et al. 2009). With the expected increase in renewable energy, careful planning is needed to avoid conflicts between the development of green energy and concerns with wildlife impacts.

   With the exception of hydropower, the two primary sources of commercial renewable energy in North America are electricity produced from wind (7 percent of U.S. renewable energy consumption in 2008) and bio-fuels (19 percent of U.S. renewable energy consumption in 2008) from food and forage crops. For comparison, U.S. solar (including photovoltaic) energy made up only 1 percent of renewable energy consumption in 2008. Technological advances are under way in the solar sector that may soon improve efficiencies and reduce costs, but for the foreseeable future, wind and biofuels will probably remain the dominant sources of renewable energy (other than hydropower) in North America.

Existing native grasslands, restored grasslands, and shrublands of the western United States support a great diversity of wildlife and provide critical ecological goods and services, such as maintenance of water quality and sequestration of atmospheric carbon. Bird diversity is especially high in unique regions such as the glaciated Prairie Pothole Region (PPR) in the midcontinent of North America, and expanses of sagebrush shrublands in the Rocky Mountains provide critical habitat for migrating ungulates and upland bird species. The way in which new demand for biofuels and wind plays out across the West will largely determine whether grassland- and shrubland-dependent populations of wildlife benefit or suffer.

In this chapter, we summarize what is known about how wind and bio-fuel energy production affect wildlife and their habitats across western North America, and discuss the extent of land potentially affected by these two sources of energy development if the United States and Canada are to meet their stated renewable energy goals. We close with recommendations to reduce wildlife impacts through proper siting, preassessment impact studies, and mitigation.

Wind Energy and Wildlife Conservation

Commercial wind energy facilities have been constructed in thirty-five U.S. states (American Wind Energy Association [AWEA] 2009), and total wind power capacity in the United States increased from 10 megawatts of nameplate capacity in 1981 to 29,440 megawatts as of June 2009 (AWEA 2009), enough to provide electricity to 8 million average households. Despite rapid growth, wind energy amounted to less than 1 percent of U.S. electricity generation in 2006. The wind industry is planning to generate 6 percent of the country's electricity supply by 2020 (AWEA 2006). Canada has 2,854 megawatts of installed wind capacity, meeting about 1 percent of Canada's demand (Canadian Wind Energy Association 2009).

Wind energy has the potential to reduce environmental impacts caused by fossil fuels because wind power does not generate atmospheric contaminants or thermal pollution (National Academy of Sciences [NAS] 2008). Therefore, wind-generated electricity does not have many of the negative environmental impacts associated with other energy sources, such as air and water pollution, or greenhouse gas emissions associated with climate change (Arnett et al. 2007). Wind power is a domestic source of energy that can be produced without water consumption, mining, drilling, refining, waste storage, or many of the other problems that accompany traditional forms of energy generation (Federal Advisory Committee 2009).

Although wind energy has many environmental benefits, wind energy development has caused the deaths of birds and bats that collide with turbines and has resulted in indirect impacts to wildlife through behavioral displacement and habitat loss (Arnett et al. 2007; NAS 2008). We define western North America as the Canadian provinces of Manitoba, Saskatchewan, Alberta, and British Columbia and the states of North and South Dakota, Nebraska, Kansas, Oklahoma, Texas, Montana, Wyoming, Colorado, New Mexico, Idaho, Utah, Arizona, Nevada, Washington, Oregon, and California. Although we limit our review to these states and provinces, some relevant data collected in other states (e.g., Minnesota) were included. Of the seventeen western U.S. states, only Nevada and Arizona did not have any installed wind energy as of June 27, 2009 (AWEA 2009). The remaining fifteen states had 19,951 megawatts of installed capacity, which represents 68 percent of all installed wind energy in the United States. Texas, with 8,361 megawatts, is by far the largest, followed by California, with 2,787 megawatts. The amount of installed capacity in the other western states ranges from 20 megawatts in Utah to 1,575 megawatts in Washington (AWEA 2009). Of the western Canadian provinces, Alberta has the largest installed capacity (524 megawatts), followed by Saskatchewan (171 megawatts) and Manitoba (104 megawatts) (Canadian Wind Energy Association 2009). Unlike several wind energy facilities in the eastern United States, all wind energy facilities in western North America are located in nonforested habitats, including agricultural fields, grasslands, and shrub-steppe.

Collision Mortality

Concerns about avian collision mortality at wind energy facilities originated when high raptor fatality rates were first reported at the Altamont

Pass Wind Resource Area (APWRA) in California (Orloff and Flannery 1992). An estimated 881–1,300 raptors are killed annually in the APWRA, which equates to 1.5–2.2 raptor fatalities per megawatt per year, the most common being golden eagles (*Aquila chrysaetos*), red-tailed hawks (*Buteo jamaicensis*), American kestrels (*Falco sparverius*), and burrowing owls (*Athene cunicularia*) (Smallwood and Thelander 2004). The APWRA consists primarily of small, older-generation turbines, many of which have lattice support towers, and many of the electrical lines are above ground, providing opportunities for raptors to perch throughout the facility. There are currently more than 5,000 turbines of various types and sizes with an installed capacity of 550 megawatts. Most of the turbines range in size from 40 to 300 kilowatts, the most common size being 100 kilowatts (Arnett et al. 2007). The two other large, older-generation wind energy facilities in California (San Gorgonio and Tehachapi) have not experienced the level of raptor fatalities observed at the APWRA (Anderson et al. 2004, 2005). Differences in raptor fatality rates between these three sites appear to be related to raptor densities.

More recent wind energy developments consist of much larger turbines, ranging in size from 0.66 to 2.5 megawatts, with tubular steel towers and three-blade rotors; most electrical lines are buried. Raptor fatality rates at facilities with modern turbines in western North America have generally been much lower than at the APWRA. At eighteen modern facilities in western North America where raptor fatality estimates are available, raptor fatality rates have ranged from 0 to 1.79 per megawatt per year and averaged 0.19 per megawatt per year (table 8.1).

The two facilities with the highest raptor fatality rates (1.79 and 0.53 per megawatt per year) are in California. Of the sixteen facilities located outside California where raptor fatality rates were reported, raptor fatality rates have ranged from 0 to 0.15 and averaged 0.07 per megawatt per year, or approximately seven raptors for each 100 megawatts of development. These facilities include nine located in Washington and Oregon, three in Alberta, and one each in Montana, Wyoming, Nebraska, and Texas. Although raptor fatality rates are generally low at most modern wind energy facilities, the number of raptor fatalities is still much higher than that of passerines (i.e., songbirds) relative to the number of individuals exposed to collisions (NAS 2008).

Mortality estimates for all bird species combined are publicly available for twenty-one wind energy facilities in western North America, including those mentioned earlier plus two facilities in Oklahoma and one additional California facility. Bird fatality rates have ranged from 0.08 to 5.67 per

TABLE 8.1. Avian and bat fatality rates at modern wind energy facilities in western North America.

Study Area	Raptor Fatalities (per MW/year)	Bird Fatalities (per MW/year)	Bat Fatalities (per MW/year)	Total MW	Reference
Diablo Winds, CA	1.79	5.67	0.78	20.0	Smallwood and Karas (2009)
Judith Gap, MT	0.09	3.01	8.93	135.0	TRC Environmental Corporation (2008)
Blue Canyon II, OK	–	0.38	3.71	151.2	Burba et al. (2008)
Nine Canyon, WA	0.05	2.76	2.47	48.0	Erickson et al. (2003)
Foote Creek Rim, WY	0.05	2.50	2.23	41.4	Young et al. (2003)
High Winds, CA	–	1.36	2.02	162.0	Kerlinger et al. (2006)
Big Horn, WA	0.15	2.54	1.90	199.5	Kronner et al. (2008)
Combine Hills, OR	0.00	2.56	1.88	41.0	Young et al. (2006)
Stateline, WA/OR	0.09	2.92	1.70	300.0	Erickson et al. (2004)
NPPD Ainsworth, NE	0.06	1.63	1.16	59.4	Derby et al. (2007)
Vansycle, OR	0.00	0.95	1.12	24.9	Erickson et al. (2000)
Klondike, OR	0.00	0.95	0.77	24.0	Johnson et al. (2003)
Hopkins Ridge, WA	0.14	1.23	0.63	150.0	Young et al. (2007)
Oklahoma Wind Energy Center, OK	–	0.08	0.53	102.0	Piorkowski (2006)
Klondike II, OR	0.11	3.14	0.41	75.0	Northwest Wildlife Consultants, Incorporated, and Western EcoSystems Technology, Incorporated (2007)

TABLE 8.1. Continued

Study Area	Raptor Fatalities (per MW/year)	Bird Fatalities (per MW/year)	Bat Fatalities (per MW/year)	Total MW	Reference
Wildhorse, WA	0.09	1.55	0.39	229.0	Erickson et al. (2008)
Buffalo Gap, TX	0.10	1.32	0.10	134.0	Tierney (2007)
SMUD, CA	0.53	0.99	0.07	15.0	URS et al. (2005)
Castle River, AB	0.01	0.29	0.84	40.0	Brown and Hamilton (2002)
McBride Lake, AB	0.09	0.55	0.71	173.0	Brown and Hamilton (2006)
Summerview, AB	0.11	1.06	12.41	70.0	Baerwald (2008), Baerwald et al. (2009)

megawatt per year and averaged 1.78 per megawatt per year (table 8.1). Avian mortality in western North America is lower than the U.S. national average. Using mortality data from a 10-year period from wind energy facilities throughout the entire United States, the average number of bird collision fatalities is 3.1 per megawatt per year (National Wind Coordinating Committee 2004).

Based on data from twenty-one fatality monitoring studies conducted in western North America at modern wind energy facilities (table 8.1), where 1,247 avian fatalities representing 128 species were reported, raptor fatalities made up 19.4 percent of the identified wind energy facility–related fatalities. The most common raptor fatalities were American kestrel (eighty-two fatalities), red-tailed hawk (forty-six), turkey vulture (*Cathartes aura*; forty-two), and burrowing owl (thirteen). Passerines were the most common collision victims, making up 59.3 percent of the fatalities, with horned lark (*Eremophila alpestris*; 272 fatalities), golden-crowned kinglet (*Regulus satrapa*; forty-seven), and western meadowlark (*Sturnella neglecta*; forty-five) experiencing the highest numbers of fatalities. Upland gamebirds were the third most common group found, making up 9.6 percent of the fatalities. Ring-necked pheasant (*Phasianus colchicus*; forty-five fatalities), gray partridge (*Perdix perdix*; thirty-eight), and chukar (*Alectoris chukar*; eighteen) were the most common fatalities found. Mourning doves (*Zenaida macroura*; twenty-nine fatalities) and rock pigeons (*Columba livia*; seventeen) made up 3.8 percent. Waterbirds such as American coot (*Fulica americana*; ten fatalities) and western grebe (*Aechmophorus occidentalis*; seven) were uncommon, representing 4.0 percent of all fatalities. Waterfowl, primarily mallard (*Anas platyrhynchos*; nine fatalities), were also infrequently found (1.9 percent of all fatalities). Only three shorebirds (0.2 percent of all fatalities) were found. Other groups, such as nighthawks, woodpeckers, and swifts, combined accounted for 1.9 percent of all fatalities. Birds that could not be identified to any avian group also made up 1.9 percent of reported fatalities.

Bat collision mortality at wind energy facilities appears to occur worldwide, as it has been documented in Australia, Germany, Sweden, Spain, and Canada (Johnson 2005). The highest bat fatality rates have occurred at facilities located on forested ridges in the eastern United States (14.9–53.3 per megawatt per year), although high bat fatality rates have also occurred in the Northeast and Upper Midwest and in southern Alberta, suggesting that wind energy facilities located in nonforested areas may also have high bat fatality rates (Arnett et al. 2007). Bat fatality estimates are available for twenty-one wind energy facilities located throughout western North

America, where bat fatalities have ranged from 0.07 per megawatt per year at a wind energy facility in California to 12.41 per megawatt per year over a 3-year period at a facility in Alberta, and averaged 2.13 fatalities per megawatt per year (table 8.1), which is slightly higher than avian mortality at wind energy facilities in western North America.

Although it has been assumed that most bat fatalities at wind energy facilities were caused by blunt trauma, based on necropsy results of bats found in Alberta, it was determined that 90 percent of bat fatalities had internal lung hemorrhaging consistent with barotrauma, and it was hypothesized that direct contact with turbine blades accounted for only half of the fatalities (Baerwald et al. 2008). The barotraumas were assumed to be caused by rapid air pressure reduction near moving turbine blades.

Most of the mortality throughout North America occurred among migratory tree-roosting bats, namely the eastern red bat (*Lasiurus borealis*), hoary bat (*Lasiurus cinereus*), and silver-haired bat (*Lasionycteris noctivagans*) (Johnson 2005; Arnett et al. 2008). Of 2,343 bat fatalities reported from studies conducted in western North America, of which 2,285 were identified, hoary bat made up 55.9 percent, silver-haired bat made up 33.1 percent, and Brazilian free-tailed bat (*Tadarida brasiliensis*) made up 6.8 percent. Species that made up less than 2 percent each of the identified fatalities included little brown bat (*Myotis lucifugus*), big brown bat (*Eptesicus fuscus*), eastern red bat, western red bat (*Lasiurus blossevillii*), and evening bat (*Nycticeius humeralis*). Approximately 90 percent of bat fatalities occur from mid-July through the end of September, with more than 50 percent occurring in August (Johnson 2005). At most sites, mortality during the spring migration and breeding season is much lower. However, as wind energy expands into the southwestern United States, where large populations of Brazilian free-tailed bat occur, impacts may occur in breeding bat populations. Studies at facilities located in Oklahoma and California have found that this species made up 41.3 percent and 85.6 percent of the bat fatalities, respectively (Arnett et al. 2008), although total bat mortality at both of these facilities was low (2.02 per megawatt per year in California and 0.53 per megawatt per year in Oklahoma; table 8.1). Many of the free-tailed bat fatalities in Oklahoma involved breeding bats, rather than migrants, unlike most facilities in western North America (Johnson 2005; Arnett et al. 2008). Most bat mortality occurs during low wind speeds at night (Arnett et al. 2008). At wind energy facilities where bat mortality is high, curtailment of turbines during these low-wind situations has been shown to reduce bat mortality by 50–70 percent at a site in Alberta (Baerwald et al. 2009) and by 53–87 percent at a site in Pennsylvania (Arnett et al. 2009).

Although collision mortality is well documented at most wind energy facilities, population-level effects have not been detected, although few studies have addressed this issue. Available data from wind energy facilities suggest that fatalities of passerines from turbine strikes generally are not significant at the population level, although exceptions could occur if facilities are sited in areas where migrating birds or rare species are concentrated (Arnett et al. 2007). In such situations, Desholm (2009) developed a framework for ranking bird species in terms of the relative sensitivity to turbine collisions based on relative abundance and demographic sensitivity (e.g., survival rates). The framework allows preconstruction identification of the species that are at high risk of being adversely affected. When this framework was applied to an offshore wind project in Denmark, results suggested that raptors and waterbirds had the highest risk of being affected, and passerines showed low risk.

Johnson and Erickson (2008) examined the potential for population-level impacts caused by avian collision mortality associated with 6,700 megawatts of existing and proposed wind energy development in the Columbia Plateau Ecoregion of eastern Oregon and Washington. The number and species composition of bird collision fatalities were estimated based on results of eleven existing mortality studies in the ecoregion. Estimated breeding population sizes were available for most birds in the ecoregion based on Breeding Bird Survey data. Predicted mortality rates for avian groups and species of concern were compared with published annual mortality rates. Because the additional wind energy–associated mortality was found to make up only a small fraction of existing mortality rates, it was concluded that population-level impacts would not be expected for the ecoregion as a whole, but that local impacts to some species could occur. In the only study to quantitatively assess potential population-level impacts, Hunt (2002) conducted a 4-year radio telemetry study of golden eagles at the APWRA and found that the resident golden eagle population appeared to be self-sustaining despite sustaining high levels of fatalities, but the effect of these fatalities on eagle populations wintering within and adjacent to the APWRA was unknown. Additional research conducted in 2005 by Hunt and Hunt (2006) found that all fifty-eight territories occupied by golden eagle pairs in the APWRA in 2000 remained active in 2005.

Bats are long-lived species with low reproductive rates. Therefore, their populations are much slower to recover from large fatality events than other species, such as most birds, that have much higher reproductive rates (Kunz et al. 2007b). Because migratory tree bats are primarily solitary tree dwellers that do not hibernate, it has not been possible to develop any suitable field methods to estimate their population sizes (Carter et al. 2003). As a result, impacts on these bat species caused by wind energy development cannot be put into perspective from a population impact standpoint. Based on their estimates of cumulative bat impacts from wind energy development in the eastern United States, Kunz et al. (2007b) concluded that wind energy development could have a substantial impact on bat populations, especially given that evidence suggests the eastern red bat, and perhaps other species, are in decline throughout much of their range (Carter et al. 2003). Although bat mortality at most wind energy facilities in western North America is lower than in other portions of the United States (Johnson 2005; Arnett et al. 2008), the potential for causing significant population-level impacts cannot be determined without estimates of population sizes. To help solve this problem, population genetic analyses of DNA sequence and microsatellite data are being conducted to provide population size estimates, to determine whether populations are growing or declining, and to see whether these populations consist of one large population or several discrete subpopulations that use spatially segregated migration routes (Amy L. Russell, Grand Valley State University, personal communication).

Indirect Effects

In addition to direct effects through collision mortality, wind energy development results in direct loss of habitat where infrastructure is placed and indirect loss of habitat through behavioral avoidance and habitat fragmentation. Direct loss of habitat associated with wind energy development is minor for most species compared with most other forms of energy development. Although wind energy facilities can cover substantial areas, the permanent footprint of wind energy facilities, such as turbines, access roads, maintenance buildings, substations, and overhead transmission lines, generally occupies only 5–10 percent of the entire development area (Bureau of Land Management 2005). Estimates of temporary construction impacts range from 0.2 to 1.0 hectares (0.5–2.5 acres) per turbine (AWEA 2009). However, behavioral avoidance may render much larger areas unsuitable or less suitable for some species of wildlife, depending on how far each species is displaced from wind energy facilities. Based on some studies in Europe, displacement effects associated with wind energy were thought to have a greater impact on birds than collision mortality (Gill et al. 1996). The greatest concern with displacement impacts for wind energy facilities in western North America has occurred where these facilities are constructed in native habitats such as grasslands or shrublands (Leddy et al. 1999; Mabey and Paul 2007).

Most studies on raptor displacement at wind energy facilities indicate that effects appear to be negligible. A before–after control–impact (BACI) study of avian use at the Buffalo Ridge wind energy facility in Minnesota found evidence of northern harriers (*Circus cyaneus*) avoiding turbines on both a small scale (less than 100 meters [328 feet] from turbines) and a larger scale (105–5,364 meters [345–17,598 feet]) in the year after construction (Johnson et al. 2000a). Two years after construction, however, no large-scale displacement of northern harriers was detected.

The only published report of avoidance of wind turbines by nesting raptors occurred at the Buffalo Ridge facility, where raptor nest density on 261.6 square kilometers (101 square miles) of land surrounding the facility was 5.94 nests per 101.0 square kilometers (39 square miles), yet no nests were present in the 31.1-square-kilometer (12-square-mile) facility itself, even though habitat was similar (Usgaard et al. 1997). At a wind energy facility in eastern Washington, extensive monitoring using helicopter flights and ground observations revealed that raptors still nested in the study area at approximately the same levels after construction, and several nests were located within 0.8 kilometer (0.5 mile) of turbines (Erickson et al. 2004). Howell and Noone (1992) found similar numbers of raptor nests before and after construction of Phase 1 of the Montezuma Hills wind energy facility in California, and anecdotal evidence indicates that raptor use of the APWRA in California may have increased since the installation of wind turbines (Orloff and Flannery 1992; AWEA 1995). At the Foote Creek Rim wind energy facility in southern Wyoming, one pair of red-tailed hawks nested within 0.5 kilometer (0.3 mile) of the turbine strings, and seven red-tailed hawk nests, one great horned owl (*Bubo virginianus*) nest, and one golden eagle nest located within 1.6 kilometers (1 mile) of the wind energy facility successfully fledged young (Johnson et al. 2000b; Western EcoSystems Technology, Inc., unpublished data). The golden eagle pair successfully nested 0.3 kilometer (0.5 mile) from the facility for three different years after the project became operational.

Studies in western North America concerning displacement of nonraptor species have concentrated on grassland passerines and waterfowl. Wind energy facility construction appears to cause small-scale local displacement of some grassland passerines, probably because the birds avoid turbine noise and maintenance activities. Construction also reduces habitat effectiveness because of the presence of access roads and large gravel pads surrounding turbines (Leddy 1996; Johnson et al. 2000a). Leddy et al. (1999) surveyed bird densities in Conservation Reserve Program (CRP) grasslands at the Buffalo Ridge wind energy facility in Minnesota and found that mean densities of ten grassland bird species were four times higher at areas located 180 meters (591 feet) from turbines than they were at grasslands nearer turbines. Johnson et al. (2000a) found reduced use of habitat within 100 meters of turbines by seven of twenty-two grassland-breeding birds after construction of the Buffalo Ridge facility. At the State-line wind energy facility in Oregon and Washington, use of areas less than 50 meters from turbines by grasshopper sparrow (*Ammodramus savan-narum*) was reduced by approximately 60 percent, with no reduction in use more than 50 meters from turbines (Erickson et al. 2004). At the Combine Hills facility in Oregon, use of areas within 150 meters of turbines by west-ern meadowlark was reduced by 86 percent, compared with a 12.6 percent reduction in use of reference areas over the same time period (Young et al. 2005a). However, horned larks showed significant increases in use of areas near turbines at both these facilities, possibly because the cleared turbine pads and access roads provided habitat preferred by this species.

Shaffer and Johnson (2008) examined displacement of grassland birds at two wind energy facilities in the northern Great Plains. Intensive transect surveys were conducted in grid cells that contained turbines and in refer-ence areas. The study focused on five species at two study sites, one in South Dakota and one in North Dakota. Based on this analysis, killdeer (*Charadrius vociferous*), western meadowlark, and chestnut-collared long-spur (*Calcarius ornatus*) did not show any avoidance of wind turbines. However, grasshopper sparrow and clay-colored sparrow (*Spizella pallida*) showed avoidance out to 200 meters (656 feet).

At the Buffalo Ridge facility, the abundance of several bird types, in-cluding shorebirds and waterfowl, was found to be significantly lower at survey plots with turbines than at reference plots without turbines (John-son et al. 2000a). The report concluded that the area of reduced use was limited primarily to areas within 100 meters of the turbines. These results are similar to those of Osborn et al. (1998), who reported that birds at Buf-falo Ridge avoided flying in areas with turbines.

Results of a long-term mountain plover (*Charadrius montanus*) moni-toring study at the Foote Creek Rim wind energy facility in Wyoming sug-gest that construction of the facility resulted in some displacement of mountain plovers. The mountain plover population was reduced during construction but has slowly increased since, although not to the same level as it was before construction. It is not known whether the initial decline was due to the presence of the wind energy facility or to regional declines in mountain plover populations. The subsequent increase may also be influenced by regional changes in mountain plover abundance. Nevertheless, some mountain plovers have apparently become habituated to the turbines, as several mountain plover nests have been located within 75 meters (246 feet) of turbines, and many of the nests were successful (Young et al. 2005a).

Breeding puddle ducks (mallard, blue-winged-teal [*Anas discors*], gadwall [*A. strepera*], northern pintail [*A. acuta*], and northern shoveler [*A. clypeata*]) were counted on wetland complexes within two wind energy facilities and on similar reference areas in North and South Dakota during the 2008 and 2009 breeding seasons (Ducks Unlimited Inc. and U.S. Fish and Wildlife Service [USFWS], unpublished data). Based on results of the surveys, breeding puddle duck abundance was not lower than expected in areas of wind energy development in 2008, but the 2009 data suggested lower pair densities in the wind developed sites. The study is continuing through 2010 to further assess the response of breeding ducks to wind energy development.

Wind Energy and Prairie Grouse

Much debate has occurred recently regarding the potential impacts of wind energy facilities on prairie grouse (*Centrocercus* and *Tympanuchus* spp.). It is currently unknown how prairie grouse, which are accustomed to a low vegetation canopy, would respond to numerous wind turbines hundreds of meters taller than the surrounding landscape. Some scientists speculate that such a skyline may displace prairie grouse hundreds of meters or even kilometers from their normal range (Manes et al. 2002; USFWS 2003; NWCC 2004). If birds are displaced, it is unknown whether, in time, local populations may become acclimated to elevated structures and return to the area. The USFWS argued that because prairie grouse evolved in habitats with little vertical structure, the placement of tall human-made structures, such as wind turbines, in occupied prairie grouse habitat may result in a decrease in habitat suitability (USFWS 2004). Several studies have shown that prairie grouse avoid other anthropogenic features, such as roads, power lines, oil and gas wells, and buildings (Robel et al. 2004; Holloran 2005; Pruett et al. 2009a, 2009b). Much of the infrastructure associated with wind energy facilities, such as power lines and roads, is common to most forms of energy development, and it is assumed that impacts would be similar.

Nevertheless, there are substantial differences between wind energy facilities and most other forms of energy development, particularly related to human activity. Although results of these studies suggest that the potential exists for wind turbines to displace prairie grouse from occupied habitat, well-designed studies examining the potential impacts of wind turbines on prairie grouse are lacking. Ongoing telemetry research being conducted by Kansas State University to examine the response of greater prairie chickens to wind energy development in Kansas, and a similar study being conducted by Western EcoSystems Technology, Inc. (Johnson et al. 2009a) on greater sage-grouse (*Centrocercus urophasianus*) response to wind energy development in Wyoming, will help address this knowledge gap.

Other than these two ongoing telemetry studies, we are aware of only three publicly available studies that examined the response of prairie grouse species to wind energy development. The Nebraska Game and Parks Commission (2009) monitored greater prairie chicken (*Tympanuchus cupido*) and sharp-tailed grouse (*T. phasianellus*) leks after construction of the thirty-six-turbine Ainsworth wind energy facility in Brown County, Nebraska. Surveys for leks were conducted 4 years after construction (2006–2009) within a 1.6- to 3.2-kilometer (1- to 2-mile) radius of the facility, an area that covered approximately 65 square kilometers (25 square miles). The numbers of leks of both species combined in the study area were thirteen, twelve, nine, and twelve in the first 4 years after construction. The number of greater prairie chickens counted on leks increased from seventy to ninety-five during the 4-year period, whereas the number of sharp-tailed grouse decreased from sixty-six to fifty-six. No preconstruction data were available on prairie grouse leks near the site; however, densities of lekking grouse on the study area at the Ainsworth facility were within the range of expected grouse densities in similar habitats in Brown County and the adjacent Rock County (Nebraska Game and Parks Commission 2009). The leks ranged from 0.7 to 2.7 kilometers (0.4 to 1.7 miles) from the nearest turbine, with an average distance of 1.4 kilometers (0.9 miles).

At a three-turbine wind energy facility in Minnesota, six active greater prairie chicken leks were located within 3.2 kilometers of turbines, with the nearest lek within 1 kilometer (0.6 mile) of the nearest turbine (USFWS 2004). During subsequent research at this facility based on forty nest locations, it was found that nesting hens were not avoiding turbines. Based on extensive research of the prairie chicken population in the vicinity of this wind energy facility from 1997 to 2009, it was concluded that the distribution and location of leks and especially nests were determined by the presence of adequate habitat in the form of residual grass cover, not the presence of vertical structures such as trees, woodlots, power lines, and wind turbines (Toepfer and Vodehnal 2009).

Greater prairie chicken lek surveys were conducted at the Elk River wind energy facility in Butler County, Kansas, within the southern Flint Hills, beginning 3 years before and continuing 4 years after construction (Johnson et al. 2009a). The facility consists of 100 1.5-megawatt turbines. During the year immediately before construction of the project (2005), ten leks were present on the project area, with 103 birds on all leks combined. By 2009, 4 years after construction, only one of these ten leks remained active, with three birds on the lek. The ten leks were located 88–1,470 meters from the nearest turbine, with a mean distance of 587 meters; eight of the ten leks were located within 0.8 kilometer (0.5 mile) of the nearest turbine. Although this decline may be attributable to development of the wind energy facility, greater prairie chicken populations have declined significantly in the Flint Hills due to the practice of annual spring burning. During the same time frame that leks were monitored at the Elk River facility, the estimated average number of greater prairie chickens in the southern Flint Hills declined by 65 percent from 2003 to 2009. In Butler County, the estimated number of birds declined by 67 percent from 2003 to 2009 (Kansas Department of Wildlife and Parks, unpublished data). This regional decline is attributed primarily to the practice of annual spring burning and heavy cattle stocking rates, which remove nesting and brood-rearing cover for prairie chickens (Robbins et al. 2002). Therefore, it seems unlikely that the decline of prairie chickens on the Elk River site is due entirely to the presence of wind turbines, although the presence of turbines may have contributed to the decline (Johnson et al. 2009a).

Another grouse with a lek mating system, the black grouse (*Lyrurus tetrix*), was found to be negatively affected by wind power development in Austria (Zeiler and Grünschachner-Berger 2009). The number of displaying males in the wind power development area increased from twenty-three to forty-one during the 3-year period immediately before construction but then declined to nine males 4 years after construction. In addition to the decline in displaying males, the remaining birds shifted their distribution away from the turbines. One lek located within 200 meters of the nearest turbine declined from twelve birds 1 year before construction to no birds 4 years after construction.

Although the data collected in the United States on response of prairie grouse to wind energy development indicate that prairie grouse may continue to use habitats near wind energy facilities, research conducted on greater sage-grouse response to oil and gas development has found that population declines due to oil and gas development may not occur until 4 years after construction (Naugle et al. 2011). Therefore, data spanning two or more grouse generations will be needed to adequately assess the impacts of wind energy development on prairie grouse.

There is little information regarding wind energy facility operation effects on big game. At the Foote Creek Rim wind energy facility, pronghorn antelope (*Antilocapra americana*) observed during raptor use surveys were recorded year-round (Johnson et al. 2000b). The mean number of pronghorn antelope observed at the six survey points was 1.07 per survey before construction of the wind energy facility and 1.59 and 1.14 per survey the 2 years immediately after construction, indicating no reduction in use of the immediate area. During a study of interactions of a transplanted elk (*Cervus elaphus*) herd with operating wind energy facilities in Oklahoma, no evidence was found that operating wind turbines have a measurable impact on elk use of the surrounding area (Walter et al. 2009). Current telemetry studies being conducted to assess the response of elk to wind energy development in Wyoming and Oregon and pronghorn antelope response in Wyoming will help address potential impacts on big game.

Planning for Wind Energy

Meeting the goal of 20 percent wind energy by 2030 would require use of an area of approximately 50,000 square kilometers (19,300 square miles) for land-based wind energy facilities and more than 11,000 square kilometers (4,245 square miles) for offshore facilities (NAS 2008), although the direct loss of land would make up only 5–10 percent of this amount (Bureau of Land Management 2005). McDonald et al. (2009) examined energy development impacts on habitat under various cap-and-trade scenarios and ranked the expected wind energy development footprint behind only that of biofuels, with 72.1 square kilometers (28 square miles) needed for each terawatt-hour produced per year by 2030. As wind energy expands across the region, the potential for cumulative impacts increases. Major expansion of renewable energy is still in its infancy, so the opportunity to avoid earlier mistakes associated with siting of oil and gas developments, as well as other forms of development in western North America, is still possible. Many government agencies and several nongovernment organizations (NGOs) have developed resource maps and recommendations to help guide future wind energy development to reduce impacts on wildlife and sensitive habitats at state (e.g., Martin et al. 2009) and national levels, such as the mapping effort being conducted jointly by the American Wind and Wildlife Institute and The Nature Conservancy (American Wind and Wildlife Institute 2009). The USFWS, along with numerous stakeholders, including the wind energy industry, academia, and conservation organizations, through the Federal Advisory Council, are developing guidelines for siting wind energy facilities, conducting preconstruction wildlife surveys, and evaluating the impacts of the facilities.

Wind energy development mitigation measures have been developed to prevent or reduce impacts and to compensate for impacts through off-site habitat mitigation when warranted (Johnson et al. 2007). Because environmental impacts determined on a project-by-project basis can underestimate the cumulative impacts of many projects, Kiesecker et al. (2009) have developed a framework to show how combining landscape-level conservation planning with application of a mitigation hierarchy can be used to determine where impacts on biodiversity can be offset and where they should be avoided or reduced (chap. 9). Use of these planning tools, guidelines, and appropriate mitigation strategies will facilitate sustainable development of wind energy in western North America, while reducing associated impacts on wildlife resources.

While wind farms are cropping up throughout the high plains and the far West, biofuels are farmed predominantly farther east (e.g., North Dakota, South Dakota, and Kansas). This geographic separation nearly ensures that all western communities are touched by some, if not many, forms of renewable energy production.

Biofuels: Boon or Bane for Grassland Wildlife?

As a rule, change comes slowly to the agricultural landscape of the West. Change can be risky, and farmers are risk averse by nature. Modern agribusiness is characterized by huge capital investment, large volumes of money moving into and then back out of checkbooks, and—in the end—thin profit margins. There is little room for experimentation and a failed crop year. But a daring new attitude is spreading across farm country, and it is setting the stage for dramatic changes in land use. Perhaps this is best characterized as the twenty-first-century gold rush of ethanol production.

Companies have been fermenting corn into ethanol for decades, and many of us have filled our vehicles with a 10 percent blend of their product. But until recently, few of us have given serious thought to the need to wean the United States off foreign oil. And only a few botanists were familiar with a plant called switchgrass that, given time and new technology, might be an important ingredient in the recipe for energy independence.

How quickly things change. Reducing our dependence on foreign oil and reducing greenhouse gas emissions are now cornerstones of federal policy. And the obscure plant called switchgrass emerged in the public consciousness when former President Bush referenced it in his 2006 State of the Union address. From all indications, biofuels are here to stay.

Energy from Biomass

Biofuels are energy sources derived from plants or other living organisms (i.e., biomass). Converting plant matter to biofuel can be as simple as picking the stems and leaves from a field, bundling them up, and transporting them to a power plant, where they are burned—together with coal and other fuels—to generate electricity. But most often, the term *biofuel* refers to a liquid transportation fuel derived from plant material. Ethanol and biodiesel are the most common examples.

Energy technology is evolving quickly, and the list of biofuels and useful coproducts continues to grow. Today, one biofuel—ethanol—is the focus of attention. Fermenting ethanol from corn or other grains is a proven technology, and engines have already been adapted to burn gasoline–ethanol blends. Currently, there are 170 corn ethanol plants in the United States, and 24 more are being built or are planned. At a conversion rate of 10.6 liters (2.8 gallons) of ethanol per bushel of corn, producing the 39.9 billion liters (10.5 billion gallons) of ethanol flowing from today's plants takes nearly 3.8 billion bushels. Corn acreage in the United States rose by 5.5 million hectares (13.5 million acres) in 2007 to nearly 37.6 million hectares (93 million acres) (National Agricultural Statistics Service 2007), a level of production that has not been seen since 1944 (fig. 8.1).

This is probably not an anomaly but rather a trend, because if all the new plants are built, corn ethanol production will reach 47.6 billion liters (12.6 billion gallons) and probably continue to increase. In December 2007, former President Bush signed the Energy Independence and Security Act of 2007, which established a new Renewable Fuels Standard of 136.8 billion liters (36 billion gallons) per year by 2022. Fifty-seven billion liters (15 billion gallons) are expected to come from corn ethanol. And the U.S. Environmental Protection Agency (2009) is considering increasing the allowable ethanol blend content from the current 10 percent to 15 percent.

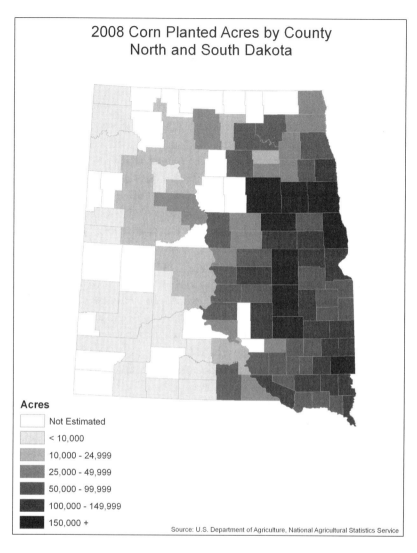

FIGURE 8.1. Distribution of corn-planted acres by county in North Dakota and South Dakota, 2008. As new varieties of corn are developed, the distribution of corn acres is likely to shift northwest from Minnesota and Iowa into central North and South Dakota. Additional tillage for corn will place our highest wetland density landscapes at increased risk of wetland drainage (fig. 8.2). (From U.S. Department of Agriculture, National Agricultural Statistics Service.)

All of this is moving the United States toward the "30 by 30" goal of replacing 30 percent of transportation fuel consumption with renewable fuels by the year 2030.

But consider that these old and new plants, together with plant expansions, will have a combined need for 4.5 billion bushels of corn. That number represents 37 percent of the 2008 U.S. corn harvest (fig. 8.1). This new demand will cut into the corn supply that is already being used for animal feed and as a key ingredient in thousands of food products, most notably processed food. Speculation continues to mount that someday soon there may not be enough corn to go around. Recent passage of the Energy Independence and Security Act of 2007 will ease that speculation, but it is unclear for how long. There is little doubt that corn supplies will be stretched thin, and a new dimension has been added to the ethanol challenge.

Ethanol production goals in the United States cannot be achieved solely with corn and other starch grains. Achieving these goals will require the use of new sources of biomass, such as switchgrass, wheat straw, and corn stover (i.e., cornstalks, cobs, and leaves). In each case, the portion of the plant digested to make ethanol is the cellulose—the material that provides structural support for plant growth. Ethanol that results from the digestion process is called cellulosic ethanol. It is created using specialized enzymes and a series of complex processes.

In the near future, corn and cellulosic ethanol feedstocks may well be competing for the same acres of farmland. This includes land that is now growing wheat, beef, or—in the case of land enrolled in the CRP—wildlife. In the grasslands of the West, big changes are coming. The key question is the net effect on wetlands, grasslands, and wildlife.

Current Challenges

Most experts believe that commercial-scale cellulosic ethanol processing plants will not come online for several years. However, some progress is being made, with twenty-three cellulosic plants under development or construction in the United States as of early 2009 (Renewable Fuels Association 2009). Until then, corn will remain the primary product market and demand for corn ethanol will increase, as indicated by the passage of a new Renewable Fuels Standard. Regrettably, when it comes to wildlife habitats, it is hard to find a silver lining in a forecast that calls for growing more corn. In fact, the demand for more corn is likely to increase conversion pressure on existing grasslands and wetlands throughout the PPR (fig. 8.2).

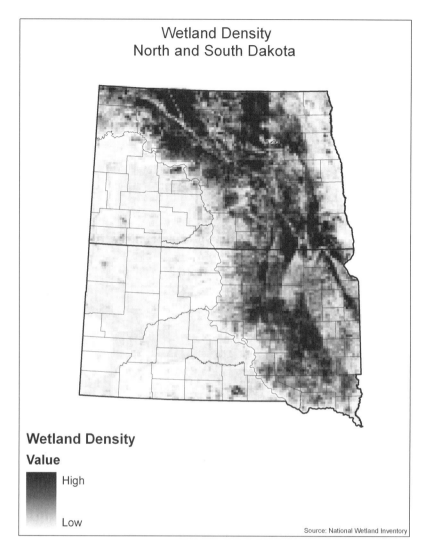

FIGURE 8.2. The highest wetland and waterfowl densities in the United States occur in the central Dakotas, an area that currently contains few harvest acres of corn (fig. 8.1). But as new varieties of corn are developed, the distribution of corn acres is likely to shift northwest from Minnesota and Iowa into the heart of the Prairie Pothole Region (wetland density map courtesy of U.S. Fish and Wildlife Service National Wetland Inventory).

Typical of any supply-and-demand relationship, when the demand for corn is high, so is its price. If corn growers receive a high price for their product, they can then afford to invest more in land and land rental (fig. 8.1). Simple supply-and-demand economics force those currently receiving a low return on their land investment to reconsider how their land can best be used to generate revenue. More area planted to corn means less area planted to other commodities, such as wheat and soybeans. Lower production of these commodities leads to dwindling supplies and increased prices. It is a vicious cycle and a situation that places an enormous amount of conversion pressure on the grasslands and wetlands of the PPR that has not been seen for decades.

An obvious concern associated with increased corn production is the fate of land enrolled in CRP, a program that was created in the 1985 Farm Bill to idle highly erodible land by restoring it to grassland. As the many other conservation benefits of CRP emerged—including the annual addition of 2.1 million ducks to the fall flight (Reynolds et al. 2006)—the program became recognized as the most significant and successful conservation initiative ever implemented by the U.S. Department of Agriculture. The CRP has been particularly important for wetlands in the PPR (fig. 8.2), with more than 1.2 million acres enrolled in various wetland practices in North and South Dakota (U.S. Department of Agriculture, Farm Services Agency, Economic and Policy Analysis Staff, September 2007). Now, those in search of more corn ground are viewing land enrolled in CRP as a reserve of cropland waiting to be brought back into production. The U.S. Department of Agriculture supported this position when it announced that there would be no new general CRP signups in 2007 or 2008—all for the stated purpose of providing more area to meet the demand for corn. Unfortunately, much area in the PPR will expire between 2007 and 2012, threatening hundreds of thousands of hectares of wetlands. Agricultural producers chose not to re-enroll 271,000 hectares (670,000 acres) across North and South Dakota that expired September 30, 2007. Two-thirds or 1.4 million hectares (3.4 million acres) of CRP grasslands in North and South Dakota will expire by 2012. This is bad news for PPR wetlands. Wetland regulations do not apply to wetlands restored as a requirement of the CRP contract, and therefore they are eligible for drainage once the contract expires. Wetlands that do remain have limited habitat value when surrounded by corn and are degraded by the impacts of sedimentation, pesticides, herbicides, and fertilizer (Gleason and Euliss 1998).

The demand for corn has also put pressure on the 9 million hectares (22.3 million acres) of remaining native prairie grassland in the U.S. portion of the PPR, which also encompasses our best remaining wetland landscapes (fig. 8.2). Most existing native prairie is not suitable for growing a high-yield corn crop. In general, the soils are poor and the climate is often extremely dynamic, two factors that generally lead to poor corn yields. However, some newly broken prairie may be suitable for wheat and other crops that will be displaced by corn on existing cropland. And the demand for corn and cropland in general is driving up the cost of all land dramatically. These increased land values also affect the region's other producers: cattle ranchers.

The livestock industry is native prairie's reason for being in an economic sense. If that industry were to disappear today, no obvious economic return could then be realized from native prairie grasslands and wetlands. Yet history tells us that some creative mind will find a way to make money from the land, but most uses will not be as compatible as ranching is with maintaining the wetlands and grasslands.

Even though cattle prices have been good for the last few years, the economics of ranching hinge on being able to buy or rent pasture at an affordable price. Higher land prices and rental rates squeeze the rancher's bottom line. Further compounding the problem are the high costs ranchers incur when they take their livestock to the feeder, the last step in the cattle-rearing process, when the animals are fattened on an increasingly valuable commodity: corn. Cattle feeders simply pass on their higher corn prices to ranchers. If a combination of high pasture values, high feeder costs, and a collapse in cattle prices occurs, many ranchers in the PPR may be driven out of business, which could spell disaster for grasslands and wetlands.

Expansion of corn for ethanol may also have unintended and unforeseen consequences for the other critical element in the wildlife equation: wetlands. Millions of wetlands in the PPR are currently protected by a strong disincentive in the federal Farm Bill called Swampbuster. Very simply, if farmers choose to drain a wetland, they disqualify themselves from all Farm Bill programs, including commodity payments that compensate them for low yields and low crop prices. To date, this disincentive has been very effective at protecting wetlands from drainage. But its effectiveness hinges on low or volatile commodity prices and the need for farmers to manage their risk in the face of these circumstances. If prices for corn and other commodities increase and remain stable, some farmers may not need or expect to receive commodity payments. That scenario effectively removes the protection afforded wetlands, and widespread wetland drainage could be the result.

These additional pressures on grasslands and wetlands in the PPR also reinforce the need for stronger protection of existing grassland. Conservation organizations have been strong advocates for a new protection measure called Sodsaver in the new Farm Bill. The Sodsaver provision would remove eligibility for crop insurance and disaster assistance on any areas without a cropping history. Federal crop insurance and disaster payments have been important incentives to convert existing grassland to cropland (Governmental Accountability Office 2007). Protection of native grasslands also prevents degradation of the adjacent wetlands.

Future Opportunities

If the wetlands and grasslands of the PPR can weather the corn ethanol storm, positive trends may be on the horizon. If implemented in a thoughtful and environmentally friendly way, the cellulosic ethanol industry could be an asset to the PPR. The key lies in the nature of the feedstock, what land uses it displaces, and how it is harvested. A perennial grass crop such as switchgrass clearly holds the potential for benefits to wetlands and wildlife.

Switchgrass, like other perennial plants being considered as biofuels, is a tall, dense grass that makes the most of the sun's energy and the soil's nutrients. The economics of cellulosic ethanol are all about the tons of biomass that can be grown on an acre of ground. Currently, switchgrass varieties being grown in the PPR yield 10–15 tons per hectare (2–3 tons per acre) annually. However, new varieties being developed through advanced genetic technology may eventually yield as much as 50 tons per hectare (10 tons per acre) annually. From a wildlife perspective, grasses used to produce cellulosic ethanol are harvested after the growing season, in fall or winter, which is beneficial to grassland-nesting birds.

Switchgrass and other perennial grasses can also play a key role in wetland conservation, especially if they are planted on cropland that used to be cultivated every year. The addition of grassland around these wetlands will greatly improve water quality, wetland function, and wetland value for wildlife. Annually tilled crops degrade wetlands and provide poor habitat for breeding grassland birds. However, the key issue is whether switchgrass replaces existing cropland or replaces other grassland that has higher value for wildlife. Switchgrass could have negative impacts on wildlife and grassland resources, especially if native grasslands or CRP are converted to monotypic stands of switchgrass (Murray and Best 2003).

One important economic consideration for cellulosic feedstocks such as switchgrass is the transportation distance from field to processing plant. At current switchgrass yields, the maximum transport distance is about 80.5 kilometers (50 miles) (Hettenhaus 2004). So policies intended to encourage the cellulosic ethanol industry should recognize the need to cluster dedicated biomass crops near proposed or existing plants. Conservation groups have suggested that existing acres of CRP in the PPR are not a viable feedstock because they are randomly scattered across the landscape. It is estimated that most of the biomass would be too far from a processing plant to be used for fuel production.

Domestic energy needs can be met while maintaining the important natural resource values and ecological goods and services provided by grassland and wetland habitats. Detrimental effects on grassland and wetland systems will be reduced in the near term if corn acreage is expanded into existing cropland and the crops displaced do not expand into native prairie or lands now enrolled in CRP. However, recent expiration of hundreds of thousands of acres of CRP and ongoing conversion of native prairie (Stephens et al. 2008) make it clear that expansion of corn acreage is having an impact on other crops and on producer decisions. On the other hand, the production of biofuels from switchgrass or other perennial grasses could provide substantial benefits for both energy and natural resources if they replace existing cropland and are properly managed.

Conclusion

Few emerging industries have as much potential as renewable energy to shape the landscape and the habitat on which wildlife depends. Conservation, industry, and government leaders must recognize this potential and become deeply engaged in energy issues to identify opportunities to benefit our natural resources, to minimize potential adverse impacts, and to mitigate the impacts that are unavoidable. The way forward is clear enough. Where feasible and cost-effective, we need to embed renewable energy production in already disturbed areas, through solar panels on the rooftops of our cities, geothermal heating of buildings, wind turbines on agricultural and mined lands, and switchgrass and other perennial grasses planted on existing croplands. When impacts do occur, we need to apply offsite mitigation planning tools to identify the species and systems most affected and fund conservation projects that offset these effects. These changes, coupled with the widespread adoption of smart grids that conserve and monitor consumer energy use and emerging clean energy technology advances on the horizon, may finally advance renewable energy production well above the current level of 7 percent of U.S. domestic energy production without compromising our wildlife and ecosystems in the process. With these changes, we can solve the green dilemma.

PART III

Conservation by Design:
Planning and Implementing Solutions

Strategic planning and implementation of conservation in our most intact and productive landscapes is a proactive solution for reducing energy impacts on wildlife populations. Too often we waste limited resources by providing palliative care to degraded landscapes where conservation costs are high and probability of success is low. Strategic planning is a top-down approach to identifying where conservation will yield the highest biological return on investment. Implementing the right conservation actions in the right places is a bottom-up approach whereby stakeholders find local solutions to maintain rural ways of life and compatible land uses. Science support is a key ingredient in spatially identifying priority landscapes and assessing conservation outcomes. Quantifying the social and biological benefits of conservation to rural communities and wildlife populations is a proven way to build long-term support for conservation.

The real question with energy development is not whether we can save every wildlife population but rather how to facilitate responsible development and still maintain functional and connected ecosystems and associated wildlife values. Part I presented a balanced overview of the spatial overlap between competing energy and wildlife values, and part II synthesized the biological impacts of poorly planned developments. Part III takes the all-important and necessary next step in resolving conflict by offering a creative

157

and viable solution. The solution is an alternative to business as usual called Energy by Design that blends a landscape vision with the mitigation hierarchy. Energy by Design provides a framework for sustainable development by first avoiding or reducing impacts on landscapes with irreplaceable wildlife values and then ensuring that impacts are restored onsite using the best available technology and finally offsetting remaining residual impacts. Offsets presented in chapter 9 provide an opportunity to mobilize an unprecedented level of funding for conservation that becomes available if we secure our energy future.

Large-scale planning maps and models will play a key role in identifying the most important landscapes for conservation and development. Chapter 10 illustrates how geographic information systems and other technologies can be used to forecast energy development scenarios to aid in conservation design. Buildout scenarios empower decision makers to evaluate tradeoffs between conservation and development before permitting decisions are made. The path forward will require governments to facilitate development in some landscapes and forgo it in others to account for uncertainties regarding wildlife impacts and to incorporate learning into management. Chapter 11 identifies ways to remove roadblocks to adaptive management on public lands and to improve policies that govern the development of federal mineral resources. Managing adaptively would enable decision makers to test assumptions by making incremental changes to leasing provisions, and provide the flexibility to modify management as new science and other data become available.

Success in conservation ultimately depends on people's willingness to communicate and their ability to work together. In part III we champion community-based landscape conservation as the next frontier for maintaining large and intact landscapes that support rural ways of life and healthy wildlife populations. In chapter 12, conservation practitioners from the fabled Blackfoot Valley in northwest Montana offer this approach as a portable and time-tested model. This "partner-centric" and scientifically credible approach to working with people links ecology and economy with social viability. The ultimate question now is whether decision makers can muster the political and economic will to replicate the basic tenets of community-based landscape conservation in priority landscapes throughout the West.

Chapter 9

Energy by Design: Making Mitigation Work for Conservation and Development

JOSEPH M. KIESECKER, HOLLY E. COPELAND,
BRUCE A. MCKENNEY, AMY POCEWICZ,
AND KEVIN E. DOHERTY

The world faces a mass extinction event that threatens 10–30 percent of all mammal, bird, and amphibian species (Wilson 1992; Novacek and Cleland 2001; Kiesecker et al. 2004; Levin and Levin 2004). Anthropogenic stressors, such as invasive species, overexploitation, pollution, and climate change, contribute to the crisis, but habitat destruction is by far the most influential factor in this unprecedented loss of biodiversity (Vitousek et al. 1997b; Hardner and Rice 2002). Development pressures will increase dramatically if global economic growth doubles by 2030 as expected (World Bank 2007), and unprecedented investment in energy development—more than $20 trillion—will be needed to support this growth, especially in developing countries (International Energy Agency 2007). This surge in development will only accelerate habitat destruction. Thus, given the importance of economic development for improving human well-being, substantial improvement in our ability to balance development needs with environmental conservation is crucial.

Global trends in energy development are mirrored by activities in the western United States and Canada. Energy development in the United States will affect an area at least the size of Wyoming by 2030, and a substantial portion of the impact will occur in western North America as traditional fossil fuels and renewable energy sources are developed at an increasingly rapid pace (McDonald et al. 2009; chap. 2). In this chapter we focus on ways to improve mitigation, with the intent to provide solutions that will benefit both development interests and conservationists.

Environmental Impact Assessment and the Mitigation Hierarchy

Environmental impact assessment is a systematic, iterative process that examines the environmental consequences of planned developments and emphasizes prediction and prevention of environmental damage (Lawrence 2003). The mitigation of environmental impacts is thus a key stage of the environmental impact assessment process and lies at its core (Pritchard 1993). Practitioners seek to minimize impacts through application of the mitigation hierarchy: avoid, minimize, restore, or offset (Council on Environmental Quality 2000; fig. 9.1).

Traditionally, mitigation has been carried out on a project-by-project basis; specific measures are implemented to mitigate project impacts at a site, usually on or adjacent to the impact site. During the environmental review and permitting phase of project development, regulatory agencies as-

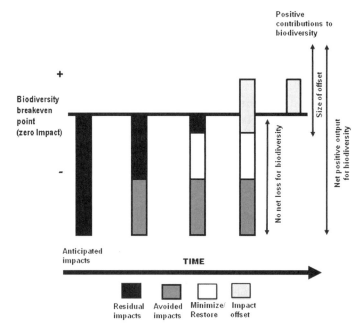

FIGURE 9.1. Achieving no net loss through application of the mitigation hierarchy.

sess the expected impacts of the project and set a proposed threshold for mitigation. The applicant or project sponsor is then responsible for developing the mitigation proposal that is presented to the agencies to confirm how project impacts can be mitigated. The mitigation can be onsite or off-site and in-kind (of similar resource or ecological function as the impact) or out-of-kind; however, there has traditionally been a preference for onsite and in-kind compensation. Project-specific mitigation is usually selected based on the impact site location, usually does not address landscape or watershed perspectives, and is generally small in scale.

Infrastructure consists of the basic facilities, such as transportation, resource extraction, utilities, and public institutions, that are needed for a functioning community or society. Often the development of infrastructure negatively affects habitat and ecosystems (Sadar et al. 1995; Canter 1996; Lawrence 2003; chaps. 3–8). Techniques have been developed to avoid, minimize, restore, and offset these impacts and the impacts of past infrastructure projects (Brown 2006). However, application of the mitigation hierarchy may not always provide the greatest environmental benefit or may do very little to promote sustainability and viability of natural systems. With conservation goals defined at a landscape level, applying the mitigation hierarchy on a project-by-project basis, often at small spatial extents, and underestimating the cumulative impacts of multiple current or future development projects undermine the hierarchy's purpose and utility. This practice does not provide a sufficient basis for determining what step in the mitigation hierarchy should be used—avoid, minimize, restore, or offset—and it limits flexibility for taking mitigation actions that would better contribute to enduring conservation outcomes. In our opinion, the way mitigation is currently applied does not capture cumulative impacts associated with development; it does not provide a structured decision-making framework to determine when projects can proceed or should be avoided; and it does not harness the full potential of offsets (conservation actions applied away from the development site).

Ours is not the first criticism of mitigation (e.g., Environmental Defense Fund 1999; Environmental Law Institute 2002; Federal Interagency Mitigation Workgroup 2002). In fact, numerous reviews (e.g., Brown 2006; Thorne et al. 2009) have identified a series of inadequacies associated with mitigation. For example, in 2001 the National Academy of Sciences and National Research Council recognized this shortcoming of traditional approaches to mitigation, identifying that often there are circumstances in which onsite or in-kind mitigation is not practicable or environmentally preferable (National Research Council 2001). Despite the recognition of the problems associated with mitigation, few attempts have been made to systematically rectify those problems. However, new regulations issued in 2008 governing wetlands (33 C.F.R. § 325 and 332; 40 C.F.R. § 230) require that compensatory mitigation for losses of aquatic resources be implemented at a watershed scale. Although the new regulations provide a sound foundation from which to develop mitigation policies, several issues warrant further policy guidance, including how best to ensure conformance with the mitigation hierarchy, identify the most environmentally preferable offsets within a landscape context, and determine appropriate mitigation replacement ratios.

For these reasons we think an overhaul of existing mitigation practices is in order to more effectively address expected development in the coming years. Fortunately, existing mitigation policy in the United States can support such changes in practices without new legislation or regulations (Wilkinson et al. 2009; McKenney and Kiesecker 2010). Here we outline a four-step framework we call Energy by Design (EbyD), developed to address the key deficiencies in mitigation and mesh with existing environmental impact assessment and regulatory approaches. Our emphasis is on blending landscape conservation planning and the mitigation hierarchy to identify situations where development plans and conservation outcomes may be in conflict and on maximizing the return provided by offsets.

1. Develop a landscape conservation plan (or use an existing conservation plan such as an ecoregional assessment). — Landscape level

2. Blend landscape planning with the mitigation hierarchy to evaluate conflicts based on vulnerability and irreplaceability.

3. Determine residual impacts associated with development and select the optimal offset portfolio. — Project level

4. Estimate offset contribution to conservation goals.

Step 1. Develop a Landscape Conservation Plan

Regulatory agencies often require developers to follow the mitigation hierarchy (Council on Environmental Quality 2000) of seeking to avoid, minimize, and restore biodiversity onsite before considering an offset for the residual impacts. However, no quantitative guidelines exist to guide this decision-making process. Therefore, a key question concerning the application of mitigation centers on when impacts from planned developments should be avoided or minimized onsite and when they should be offset (Thorne et al. 2006; Kiesecker et al. 2009). Conservation planning, particularly ecoregional assessments (e.g., Groves 2003), provides the structure to ensure that mitigation is consistent with conservation goals, which often include the maintenance of large and resilient ecosystems to support both healthy wildlife habitats and human communities. Blending the mitigation hierarchy with landscape planning offers distinct advantages over the traditional project-by-project approach because it considers the cumulative impacts of both current and projected development, provides regional context to better guide which step of the mitigation hierarchy should be applied (i.e., avoidance versus offsets), and offers increased flexibility for choosing offsets that can maximize conservation return by providing funding for the most threatened ecosystems or species.

Landscape-level conservation planning is the process of locating and configuring areas that can be managed to maintain viability of biodiversity and other natural features (Pressey and Bottrill 2008). The resulting conservation plan is intended to clearly articulate a vision that incorporates the full range of biological features, their distribution, and the minimum needs of each to persist in the long term (e.g., Lovejoy 1980; Armbruster and Lande 1993; Doncaster et al. 1996). Creation of the vision depends on the active involvement of host governments, experts of many disciplines, development organizations, and citizens in the region. A conservation portfolio of priority sites, the end product of conservation planning, is a selected set of areas that represents the full distribution and diversity of the biological features the plan attempts to conserve (e.g., Noss et al. 2002). Often plans use an optimization approach automated with spatial analysis tools to meet the minimum viability needs of each biological target yet minimize the amount of area selected (Pressey et al. 1997; Ball and Possingham 2000). Optimization tools provide flexibility for reconfiguring conservation plans in the context of future development scenarios. To date, conservation plans (i.e., ecoregional assessments) have been conducted globally, making the data needed to implement EbyD readily available (Dinerstein et al. 2000; Groves 2003).

Step 2. Blend Landscape Planning with the Mitigation Hierarchy

Energy by Design entails analysis of conservation plans in the context of future development. Proposed projects, along with projected development activities (e.g., oil and gas, wind, solar, residential, and some types of mining development), can be mapped and assessed relative to an existing conservation portfolio or incorporated in the development of a new conservation plan (Kiesecker et al. 2009). Under the EbyD approach, overlap or conflicts between a conservation portfolio and development impacts can be addressed by a redrawing of the portfolio to recapture habitat elsewhere in the study area to meet minimum viability needs of target species and ecological systems. However, if conservation goals cannot be met through redrawing of the portfolio, development impacts would need to be minimized to the degree that biodiversity values are maintained or impacts must be avoided (Kiesecker et al. 2009) (box 9.1). This approach provides an opportunity to avoid conflict between potential development and areas critical for biodiversity, as well as the structure to guide decisions regarding which step in the mitigation hierarchy should be applied in response to proposed development.

Information to support projections of future development activity can be gathered from a variety of sources, including U.S. Bureau of Land Management (BLM) resource management plans, forest management plans, long-range or metropolitan transportation plans, and community growth plans (Brown 2006). Predictive modeling approaches can be used where plans do not exist or do not provide details on the distribution of potential

---

BOX 9.1. MITIGATION PLANNING IN THE WYOMING BASINS ECOREGION

Chosen portfolio sites during the ecoregional assessment (Freilich et al. 2001) total 3.5 million hectares or 27 percent of the total area in the ecoregion. Because only 27 percent of the ecoregion was selected as part of the conservation portfolio, conservation–development conflicts could be resolved by simply redesigning the portfolio to meet target goals in areas having lower oil and gas potential. Kiesecker et al. (2010) examined the intersection between the Wyoming Basins Ecoregion conservation areas and the highest 25 percent of oil and gas potential (Copeland et al. 2009; chap. 10). Twenty-seven conservation areas intersect with areas of high development potential. Of the sites that intersect, twenty-two of twenty-seven sites could be redesigned and use offsets to mitigate impacts resulting from development. In sites where conflicts could be resolved, development could proceed with greater flexibility to apply the mitigation hierarchy, managing residual impacts through the use of onsite restoration and offsets. The five remaining sites contain highly irreplaceable targets. For these sites, greater emphasis would need to be given to avoidance or minimization.

BOX 9.1. CONTINUED

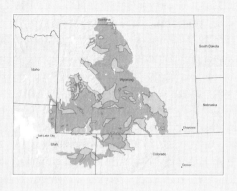

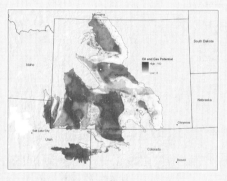

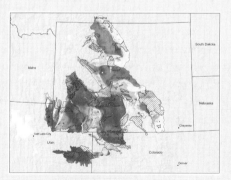

Top: Portfolio of conservation sites selected by the Wyoming Basins Ecoregional Assessment (Freilich et al. 2001). *Middle*: Oil and gas potential (Copeland et al. 2009; chap. 10). *Bottom*: Sites that overlap areas with high probability of development.

development. Although diverse predictive modeling techniques have been used in recent years to predict changes in land cover and residential development (Theobald and Hobbs 1998; Turner et al. 2007; Pocewicz et al. 2008) and to predict potential species habitat (Guisan and Zimmermann 2000; Cutler et al. 2007), only recently have similar techniques been applied to model anticipated energy development (Copeland et al. 2009; chap. 10). There is ample opportunity to use predictive modeling to describe energy development scenarios and thereby better inform decision makers and the public about patterns of anticipated development and likely impacts. When applied as part of the planning process, this approach could be used to highlight areas of biological sensitivity or avoidance areas (Kiesecker et al. 2009, 2010) necessary to achieve conservation goals for a species or to predict whether proposed development will have significant impacts on a population across its range. In this chapter we present two case studies to illustrate how the framework described earlier can be applied at a varying spatial extents focusing on a range of conservation outcomes, from individual species to a more diverse representation of biodiversity (also see chap. 10).

Mitigation Planning at an Ecoregional Scale

The Wyoming Basins Ecoregion comprises 13.3 million hectares of basin, plain, desert, and "island" mountains in Wyoming, Montana, Idaho, Colorado, and Utah (Bailey 1995; box 9.1). To identify areas that would maintain long-term persistence of representative biodiversity for the ecoregion, The Nature Conservancy and key state and federal land management and wildlife regulatory agencies, universities, and other conservation organizations set out to conduct an ecoregional plan for the Wyoming Basins Ecoregion (Freilich et al. 2001). Kiesecker et al. (2009) illustrate how the EbyD framework can be applied to the ecoregional planning approach and have applied this concept to the Wyoming Basins ecoregional assessment and energy development projected there (box 9.1).

Mitigation Planning for a Focal Species

EbyD was applied in conservation planning to evaluate options for reducing development impacts on greater sage-grouse (*Centrocercus urophasianus*) in Wyoming, Montana, Colorado, Utah, and North and South Dakota. By selecting areas with the highest population densities, managers could define core regions that contained 25, 50, 75, and 100 percent of the breeding population within 5, 12, 30, and 60 percent of the breeding range, respectively (Doherty et al. 2011). Identification and mapping of core regions provide a mechanism for assessing tradeoffs between biological value and anthropogenic risk to deliver the greatest conservation benefit to populations (Abbitt et al. 2000; Balmford et al. 2001; Wilson et al. 2005). Examining conflicts between development potential and biological value gave insight into where specific landscapes fell within the mitigation hierarchy (avoid, minimize, restore, or offset; box 9.2) and formed a frame on which the final core area plan was built by the governor of Wyoming's Sage-Grouse Implementation Task Force.

Ecological zoning of this nature is an admission that threats are large, resources are limited, and conservation action targeting every remaining population is improbable. Core regions represent a proactive attempt to

---

BOX 9.2. MITIGATION PLANNING FOR GREATER SAGE-GROUSE

Avoid and Negate Risk (Black Areas)

The simplest and most cost-effective first step in conservation is to avoid large-scale actions that further reduce or eliminate the largest populations in the best remaining landscapes. Future developments should avoid these large and intact landscapes that support core populations. Research on threshold levels of oil and gas quantify that development risk can be negated through project-level mitigation (no impacts detected less than one well per square mile) (Holloran 2005; Walker et al. 2007a; Doherty 2008).

Avoid and Offset (Darkest Gray Areas)

Areas that have high biological value and low energy potential can serve as offsets for impacts in other areas. Actions in these areas should focus on reducing risks from other stressors to sagebrush habitats such as tillage, residential development, and invasive plants such as cheatgrass.

Restore and Offset (Medium Gray Areas)

Areas of low biological value and low energy potential also represent low-conflict restoration opportunities for sage-grouse important to maintaining connectivity to high-value core regions, especially in Montana. Programs should focus on restoring adjacent lands currently in tillage agriculture to sagebrush-dominated grasslands, in addition to enhancing existing native habitats.

BOX 9.2. CONTINUED

Offset and Minimize (Lightest Gray Areas)

Areas of high energy potential and lower biological value represent areas with substantially less conflict between sage-grouse and energy development. Impacts that cannot be minimized in these areas should be used as a source of funding for restoration and offsets. Estimates of average rates of lek loss and declines in birds at remaining active leks (Doherty et al. 2010a) should be used as a currency to ensure that offsets positively influence an equal or greater number of birds.

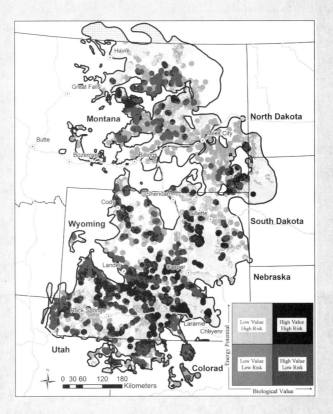

Box 9.2. Mitigation planning for greater sage-grouse.

identify and maintain a viable set of populations before the opportunity to do so is lost and direct conservation to where actions will have the largest benefit to populations.

Step 3. Determine Residual Impacts and Select the Optimal Offset Portfolio

Under the EbyD framework, proposed development consistent with conservation goals of a landscape-level conservation plan would move forward, using best-management practices to minimize impacts and harnessing offsets to mitigate for the unavoidable impacts associated with the development. Offsets are increasingly used to achieve environmental benefits (Gibbons and Lindenmayer 2007), providing a mechanism for maintaining or enhancing environmental values in situations where development is sought despite detrimental environmental impacts (ten Kate et al. 2004; McKenney and Kiesecker 2010). Offsets address residual environmental impacts of development, after efforts have been undertaken to avoid, minimize, and restore them onsite, with the overall aim of achieving a net neutral or positive outcome (fig. 9.1).

Offset policies for environmental purposes have gained attention in recent years (e.g., Environmental Defense Fund 1999; Government of New South Wales 2003; see McKenney and Kiesecker 2010 for a review). Although the use of offsets remains limited, they are increasingly used to achieve environmental benefits including pollution control, mitigation of wetland losses, and protection of endangered species (ten Kate et al. 2004). Offsets offer substantial potential benefits for industry, government, and conservation groups alike. Benefits for industry include a higher likelihood that clearance is granted from regulators for new developments, greater societal support for development projects, and the opportunity to more effectively manage environmental risks. Offsets provide government regulators with the opportunity to encourage companies to make significant contributions to conservation, particularly in situations where legislation does not require mandatory offsets. Conservation organizations can use offsets to move beyond piecemeal mitigation and thereby secure larger-scale, more effective conservation projects. Offsets can also be a mechanism ensuring that regional conservation goals are integrated into government and business planning.

Our objective with EbyD is to ensure that the use of offsets is ecologically equivalent to impacts and will persist at least as long as onsite impacts, resulting in net neutral or positive outcomes. Step 3 may in some cases be rolled into Steps 1 and 2 if planners have thought proactively about development, its impacts, and the mitigation that will be necessary. We include Step 3 here because comprehensive mitigation planning is in its infancy, and mitigation will probably occur on a project basis for some time (but see Thorne et al. 2009). Site-level mitigation planning will still need to assemble a working group, compile a list of representative biological targets, gather spatial data for biological targets, set impact goals for each biological target, and use a site-selection algorithm to identify potential offset sites and rate their landscape value and condition (Ball 2000; Ball and Possingham 2000; Possingham et al. 2000; Arponen et al. 2007).

Assemble a Working Group

Projects should always seek to apply rigorous, objective measures of conservation value while recognizing that a quantitative assessment must be supplemented by expert opinion. Stakeholder engagement is a key component of any mitigation planning exercise, and key regulatory agencies as well as members of the local community must be involved (Tisdell 1995). The advisory group that is assembled can help provide the most current spatial data on the species of concern and insights into the process being developed.

Compile a List of Representative Biological Targets

Given the challenges of measuring biological diversity directly and completely, practitioners develop a representation by selecting a set of components (targets) of biological diversity. Selecting a set of focal targets with sufficient breadth and depth has been addressed through the "coarse-filter, fine-filter" approach as applied, for example, by The Nature Conservancy's (2000) ecoregional planning. *Coarse-filter* generally refers to ecosystems; in a more practical sense, it refers to mapped units of vegetation. The idea is that conserving a sample of each distinct vegetation type, in sufficient abundance and distribution, is an efficient way to conserve the majority of biological phenomena in the target area. An oft-cited approximation is that coarse-filter conservation will conserve 80 percent of all species in a target area (Desmet and Cowling 2004). *Fine-filter* generally refers to individual species, with specific habitat needs or environmental relationships that are not adequately captured by the coarse filter. Narrow endemics and extreme habitat specialists, species with restrictive life histories, or those that have experienced significant loss of habitat or are particularly sensitive to human perturbations fall into this category (e.g., IUCN Red List species). Additional targets may also be needed to capture species of economic or social importance to local communities.

Gather Spatial Data for Biological Targets

Spatial data are critical both to quantify impacts associated with development on the project site and to guide the selection of offset sites. In cases where survey data are insufficient to estimate occurrence patterns across the study area, inductive or deductive predictive models based on species occurrence observations can be generated (Guisan and Zimmermann 2000).

Set Impact Goals for Biological Targets

Often the environmental impact assessment process will identify a footprint associated with the direct and indirect impacts associated with the projected development (Sadar et al. 1995; Canter 1996). Spatial data assembled for each of the biological targets can be overlaid onto the impact boundary, and estimated acres of habitat within these bounds can be included as impacts.

Select an Optimal Offset Portfolio

We recommend using a site-selection algorithm (e.g., Marxan; Ball and Possingham 2000; Arponen et al. 2007) to determine an appropriate location and spatial extent for offset design. Marxan is a siting tool for landscape conservation analysis that explicitly incorporates spatial design criteria into the site-selection process that can be used to maximize the value of areas selected as potential offset sites. Marxan operates as a stand-alone program and uses an algorithm called simulated annealing with iterative improvement as a heuristic method for efficiently selecting regionally representative sets of areas for biodiversity conservation (Possingham et al. 2000). Marxan allows inputs of target occurrences represented as points, lines, or polygons in a geographic information system and lets conservation goals be stated in a variety of ways, such as percentage area or numbers of point occurrences. The program also allows the integration of many available spatial datasets on land use pattern and conservation status and enables a rapid evaluation of alternative configurations. The ultimate objective is to minimize the cost of the reserve system (e.g., cost = landscape integrity, conservation cost in dollars, size of the patch) while still meeting conservation objectives. The Nature Conservancy is testing this framework through a series of pilot projects. Projects include partnerships with energy providers in the United States (e.g., British Petroleum) and collaborations with international regulatory agencies (e.g., ministries of the environment in Colombia and Mongolia). Here we present a case study to illustrate how the process is being applied.

Jonah Natural Gas Field

Located in Wyoming's Upper Green River Valley, the 24,407-hectare Jonah natural gas field is one of most substantial recent discoveries of natural gas in the United States, with an estimated 7–10 trillion cubic feet (200–300 billion cubic meters) of reserves and 500 wells in operation (U.S. Department of Interior 2006). Regulatory approval was granted by the BLM in 2006 to infill the existing 12,343-hectare developed portion of the field with an additional 3,100 wells. As a requirement of the infill project, an offsite mitigation fund of $24.5 million was established. Working with partners from state and federal regulatory agencies, universities, biological consulting firms, and the local agricultural production community, Kiesecker et al. (2009) designed an offset strategy for the Jonah Field (fig. 9.2).

Potential for Aggregated Offsets and Out-of-Kind Offsets

Landscape-level planning (Steps 1 and 2) also provides a basis for designing aggregated offsets and out-of-kind offsets. Most offset policies include a presumption for like-for-like or in-kind offsets: offsets that conserve biodiversity of a similar kind to that affected by the development. But there are situations in which better conservation results may be obtained by placing the offset in an ecosystem of higher conservation priority than that affected by the development. A regional landscape perspective may provide opportunities to identify situations in which "trading up" or out-of-kind offsets offer valuable alternatives. Consider development with impacts on a widely distributed or highly conserved target. Requiring in-kind offsets could limit the potential benefit that an offset could provide. Losses of this

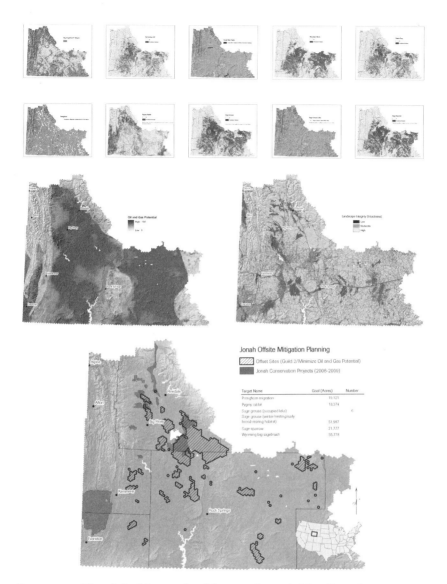

FIGURE 9.2. Use of the Marxan algorithm to select suitable offsets sites as part of the Jonah Natural Gas Field Infill Project. Spatial data layers were used for both assessing impacts as a result of development on the field and selecting suitable offset sites. "Landscape Rules: Intactness" (Copeland et al. 2007) and "Oil and Gas Potential" (Copeland et al. 2009; chap. 10) guided the selection of sites to areas of high habitat quality and low oil and gas development potential. Hatched areas represent the best fit of the Marxan algorithm based on these specific targets and specified rules. The inset map shows the location of Wyoming within the conterminous United States and the location of Wyoming Landscape Conservation Initiative within Wyoming (as modified from Kiesecker et al. 2010).

common habitat type could be offset in a habitat of higher priority in the region because it is under greater threat (is more vulnerable) or because it is the last remaining example (is irreplaceable). Out-of-kind offsets also may be preferable where there is an opportunity to increase the offset's operational feasibility by taking advantage of an existing conservation management arrangement to locate the offset, or consolidating a number of offsets in one location. Of course, alternatives to strict in-kind criteria would need to be clearly beneficial for meeting conservation objectives or adopted after proper consideration of an in-kind offset, and not simply driven by cost reduction.

Landscape-level plans also provide an opportunity to design offsets that address residual adverse impacts arising from more than one development project (Thorne et al. 2009). Aggregated offsets might be advantageous when an area is subjected to cumulative impacts from several individual developments, particularly those in the same sector, at roughly the same time. In this situation, impacts on biodiversity are likely to be of a similar type, and aggregating offsets may provide better mitigation at lower cost, with a higher probability of success given the concentration of the management skills needed to deliver the offset and synergies in project management. Such assessments can also reduce costly delays due to protracted environmental review. A landscape approach to compensatory mitigation planning can lead to a better ecological outcome. If mitigation needs from multiple projects are pooled, then larger, less fragmented parcels can be acquired, contributing to both ecological integrity and fiscal savings. There is evidence that small, isolated fragments of habitat tend to have lower overall biodiversity than larger patches (Doncaster et al. 1996).

Step 4. Estimate Offset Contribution to Conservation Goals

Offset policies generally seek no-net-loss or net-gain outcomes for conservation, compensating for impacts with offsets (McKenney and Kiesecker 2010). But how do we know when enough offset projects have been implemented to achieve no net loss? In this section we propose an accounting approach designed to improve estimates of how much an offset project compensates for project impacts, and to help identify which offsets maximize conservation return by delivering the highest-value conservation at the lowest cost and risk. Where such information on offset opportunities can be provided prospectively, it will support the selection of offsets that are both better for conservation and more cost-effective.

Under existing policies, offset benefits are often estimated using mitigation replacement ratios, which establish the number of credit units that must be debited from an offset to compensate or replace one unit of loss at the project site. Policy guidance on replacement ratios tends to fall into three categories: predefined ratios, such as those based on the type of conservation action (e.g., 1:1 ratio for restoration, 5:1 for preservation); specified assessment methods that provide a specific approach for determining ratios; and subjective determinations based on the discretion of regulatory authorities after multiple considerations, such as proposed conservation actions and risk factors (McKenney and Kiesecker 2010).

Each of these approaches has weaknesses (Kusler 2003). Although predefined ratios may simplify the offset accounting and implementation process, there is little reason to believe they deliver no-net-loss outcomes on a regular basis (King and Price 2004). Given variations in ecosystems, types of impacts, and possible offsets, predefined ratios may result in undercompensation or overcompensation, but no net loss is achieved by coincidence as much as by design. To illustrate, consider two possible restoration offsets. The first offset involves the application of an untested restoration approach, and it will be many years before conservation values are delivered, if at all. The second offset uses a well-accepted restoration approach that can deliver conservation benefits effectively and rapidly. Under offset policies using predefined replacement ratios for restoration actions, the same ratio (typically a 1:1 ratio for wetland mitigation, depending on the state) would be applied to both projects, despite the marked differences in likely conservation benefits.

Reliance on a single assessment method to determine ratios is similarly inadequate for addressing the wide range of possible impacts and offset opportunities. Consider that for wetland mitigation alone there are dozens of sophisticated assessment methods (Bartoldus 1999). These assessment methods have been developed over time in response to the contexts of different wetland types, scientific advancements, and other demands. Policies that endorse a specific assessment method for identifying replacement ratios are likely to constrain innovation and limit the potential for determining ratios that deliver no net loss. Finally, choosing replacement ratios based on the professional judgment of regulators and others is also problematic. Although this approach is more likely to consider key factors affecting replacement ratios, it is too often an ad hoc and opaque process, making it difficult to ascertain whether decisions are science-based and unbiased.

In sum, current accounting approaches are generally inadequate for ensuring no-net-loss outcomes. They are either too inflexible to address the ecological context for impacts and offsets, as is the case with predefined ratios and specified assessment methods, or too open to discretion and subjective judgment. Our proposed accounting approach seeks to strike a balance, supporting ratio determinations based on a more structured and transparent approach.

Offset Accounting

Under EbyD, a portfolio of possible offsets is identified that represents the best opportunities for protection or restoration. The portfolio selection process ensures that, in comparison to the impact site, potential offsets provide equivalent or better ecological quality, as defined by structure, condition, function, connectivity, corridors, buffers, and contribution to landscape-level conservation. Such considerations are addressed in the rule-setting process for portfolio selection.

After selection of an offset portfolio, the next question is which offsets in that portfolio should be selected for implementation. The aim is to identify offsets that will deliver the greatest contribution toward achieving no net loss at the lowest cost and risk. To assess the likely contribution of an offset to no-net-loss goals, we focus on three key elements: additionality, defined as an offset's new contribution to conservation, in addition to existing values; probability of success, defined as the likelihood that offset actions will deliver expected conservation benefits; and time lag to conservation maturity, evaluated as the length of time for offset actions to deliver conservation at a maturity level similar to what was lost at the impact site.

ADDITIONALITY

We propose an offset accounting approach that values conservation projects, whether they are restoration or preservation actions, based on their additionality. When offsets restore degraded ecosystems, they provide a new contribution to conservation over time as the offset reaches maturity. Offsets that preserve habitat also deliver conservation value when, taking into account real-world conditions and threats, those offsets protect against an expected background rate of loss. For example, protecting a 1,000-hectare forest area that was experiencing an average deforestation rate of 1 percent per year delivers a new contribution to conservation of 10 hectares per year (1 percent of 1,000 hectares). Such rates of loss can be estimated using standard threat assessments (e.g., Theobald and Hobbs 1998; Turner et al. 2007; Pocewicz et al. 2008).

PROBABILITY OF SUCCESS

Success of restoration projects can vary greatly depending on the ecosystem, restoration techniques, and other factors. In some cases, restoration approaches are known to be effective, but in other situations there may be great uncertainty due to a lack of experience. Offset accounting does not adequately take this variation into account. As noted, predefined replacement ratios are often applied regardless of the restoration approach or ecosystem. We propose an accounting approach that more effectively incorporates probability of success in the valuing of offsets. Where restoration experience is comprehensive, this probability could be estimated with some accuracy, and where experience is more limited, a high–medium–low probability ranking process might be used. Incorporating probability of success into offset accounting would ensure a more realistic appraisal of how offsets, both restoration and protection, contribute to no-net-loss outcomes. Moreover, it would create incentives for implementing effective offsets over approaches with high risks of failure and encourage offset design that includes monitoring, legal and financial assurances, adaptive management, and other measures to increase the probability of success.

TIME LAG TO CONSERVATION MATURITY

An offset that preserves habitat delivers conservation benefits at the moment it is implemented, with the level of benefits depending on the expected background rate of loss for the site. In contrast, a restoration offset may take many years before conservation benefits mature. This time lag represents a loss for biodiversity and should be accounted for in estimates of offset benefits. We propose accounting for this loss by estimating the time to maturity of a restoration action and applying a discount rate, a commonly used method for estimating the present value of future values. This approach will create appropriate incentives, making restoration offsets that promise conservation benefits far into the future less attractive than offsets that can deliver more immediate benefits. Likewise, there will be strong incentives to avoid impacts to natural systems that need very long periods to restore, as the very high replacement ratio for these restoration offsets would make them prohibitively expensive.

TABLE 9.1. Applying the proposed accounting framework.

Hectares of Impact = Goal	2,000 ha		
Offset Portfolio	Site A	Site B	Sites A and B
Hectares at offset site suitable for conservation	3,000 ha	3,000 ha	6,000 ha
Proposed conservation action	Restoration and protection	Protection	
Expected annual rate of loss of habitat	n/a	1%	
Probability of success of conservation actions	50%	100%	
Time lag to conservation maturity	30 yr	0 yr	
Offset hectare credits	1,160 ha	948 ha	2,108 ha
Percentage of goal (progress toward no-net-loss goal)	58%	47%	105%
Implicit offset-to-impact ratio	2.6:1	3.2:1	2.9:1
Cost per hectare for offset	$2,000/ha	$1,000/ha	
Total cost for offset	$6 million	$3 million	$9 million
Cost per offset hectare credit delivered	$5,172/ha	$3,165/ha	$4,270/ha

Table 9.1 illustrates in a simplified manner how our proposed accounting approach could be applied to potential offset sites within a portfolio.

In this example, we are evaluating two possible offset sites (Site A and Site B), but a typical portfolio would have many possible offset sites. The area of each site in this example is 3,000 hectares, and the goal is to deliver 2,000 hectares of offset credits to address project impacts and achieve no net loss. Whereas Site A is a degraded site proposed for restoration with protection in perpetuity, Site B is high-value habitat in good condition but in need of protection to prevent losses; the projected background rate of habitat loss is 1 percent per year (or 30 hectares per year of the 3,000-hectare site). There is no expected rate of loss at Site A because it is already degraded. The conservation benefits of the restoration offset are expected to take 30 years to reach maturity; we apply a 3 percent discount rate (National Oceanic and Atmospheric Administration 1999). The probability of success for the restoration project is estimated to be 50 percent, given uncertainties about the restoration approach, whereas success of the protection offset is assured. Based on these factors, we can estimate the offset credits that each proposed offset would deliver: 1,160 hectares for the restoration offset and 948 hectares for the protection offset. If both these offsets were selected for implementation, they would provide 2,108 hectares of credit toward the 2,000-hectares goal, and no net loss would be achieved. Rather than determining the amount of offset needed based on a predefined replacement ratio, our approach backs out the ratio based on the amount of offset credits delivered out of the total offset area (for the restoration offset, 1,160 hectares of credit out of a total site are of 3,000 hectares for a ratio of 2.6:1; table 9.1). We can also assess the likely cost to deliver this conservation benefit, as these costs are often well established. In this example, protection costs would be $1,000 per hectare or $3 million for the site, whereas the combination of restoration and protection would be $2,000 per hectare or $6 million for the site. To understand each offset's conservation return, we divide the total cost of the offset by the estimated credits it would deliver. The cost per offset credit is $5,172 per hectare for the restoration site and $3,165 per hectare for the protection site. With this information, offset implementers could identify which offsets within a portfolio would be most cost-effective to implement in order to achieve no net loss.

We recognize that developing precise estimates for the accounting framework's key factors—additionality, probability of success, and time to conservation maturity—will be challenging in some contexts. But even approximate estimates, developed using science-based methods and presented in a transparent process, would be a marked improvement over current practices for estimating compensatory mitigation. We firmly believe that offset accounting must incorporate these factors if offsets are to truly deliver on no-net-loss goals.

Fueling Conservation from Western Energy Development

Offsets represent an opportunity for mobilizing billions of dollars for conservation (Burgin 2008; McKenney and Kiesecker 2010). When offsets are a normal part of project costs, as regular practice for fully internalizing project impacts, the level of funding available for conservation can exceed other funding sources by orders of magnitude. Consider wetland mitigation, where no net loss goals and the use of wetland offsets and banks generates $3 billion in funding annually (Environmental Law Institute 2007);

compare this with federal appropriations of $60–$80 million annually for the U.S. Land and Water Conservation Fund. Likewise, in Wyoming $36 million was established as a mitigation fund for a single gas field and $24.5 million for another; compare this with the $4 million available annually for wildlife conservation from the Wyoming Wildlife and Natural Resource Trust (Kiesecker et al. 2010). The BLM, which oversees the management of more than 105.2 million hectares (260 million acres) of land in the United States and administers the mineral estate for more than 283.3 million hectares (700 million acres), has recently issued guidance on its mitigation policy supporting offsets as another tool for addressing project impacts (U.S. Department of Energy and BLM 2008). Given the extensive amount of development projected for the western United States, a requirement for development to attain no net loss could be the impetus to conserve biodiversity. This will be important as the United States moves to exploit more of its domestic energy resources, particularly renewable energy (McDonald et al. 2009).

As we look forward, the energy development footprint is likely to grow. Traditional oil and gas development will increase, with 126,000 additional oil and gas wells anticipated over the next 20 years in the Rocky Mountain West alone. This will be coupled with dramatic increases in renewable energy. For example, meeting the Department of Energy's goal of 20 percent wind energy by 2030 could result in the fragmentation of 4.8 million hectares (12 million acres) of land (U.S. Department of Energy 2008a), with 17,700 kilometers (11,000 miles) of new transmission lines to move electricity generated in areas of low population to population centers where it is needed. There are also more than 100 permit applications for solar projects pending and 14.2 million hectares (35 million acres) of economically viable solar sites identified, with potential to fragment millions of hectares of sensitive deserts in the southwest United States (U.S. Department of Interior and BLM 2008). If the mitigation framework we outline here is implemented, it would avoid losses to key wildlife resources and harness funding for conservation on a scale not seen before.

Conclusion

Balancing growing development demands with biodiversity conservation necessitates a shift from business as usual. By blending a landscape vision with the mitigation hierarchy, we move away from the traditional project-by-project approach. By first avoiding or minimizing impacts to irreplaceable occurrences of biological targets, then ensuring that impacts are restored onsite using the best available technology, and finally offsetting any remaining residual impacts, we can provide a framework truly consistent with sustainable development (Pritchard 1993; Bartelmus 1997; Clark 2007). A landscape vision is essential because it ensures that the biologically and ecologically important features remain the core conservation targets throughout the process. Without this vision, we could lose sight of overarching conservation targets, have difficulty establishing priorities, and waste scarce resources. Determining appropriate areas for habitat preservation as part of a conservation plan is a challenging exercise, but in reality this is the easy part. The real challenge is finding funding mechanisms to underwrite the conservation of these areas. By adopting the framework outlined here and requiring the application of the no-net-loss goal (Kiesecker et al. 2009), we balance development with conservation and provide the structure to fund conservation commensurate with impacts from development.

Chapter 10

Forecasting Energy Development Scenarios to Aid in Conservation Design

HOLLY E. COPELAND, KEVIN E. DOHERTY,
DAVID E. NAUGLE, AMY POCEWICZ,
AND JOSEPH M. KIESECKER

Rapid increases in development of both renewable and traditional hydrocarbon energy resources seem certain and will probably affect up to 20 percent of the major terrestrial ecosystems in the western United States (chap. 2; McDonald et al. 2009). Growing concerns over the potential social and biological impacts of climate change, with related calls to reduce carbon emissions, have intensified demands to develop renewable energy resources (Brooke 2008). Nevertheless, fossil fuels will continue as a primary source of energy, with ramifications for the western United States, where a substantial portion of domestic onshore hydrocarbon resources are found (chap. 2). Exploitation of a wider portion of our domestic energy resources will increase the likelihood of conflicts between energy development and conservation and necessitate more proactive approaches to environmental mitigation (chap. 9).

New predictive modeling techniques lend the ability to map and model both energy potential and biodiversity distributions, to better understand the scope and scale of probable impacts, and thus to address these issues before widespread development occurs. Although diverse predictive modeling techniques have been applied in recent years to project land cover changes and residential development (Theobald and Hobbs 1998; Pontius and Malanson 2005; Pocewicz et al. 2008) and to predict potential habitat for individual species (Guisan and Zimmermann 2000; Phillips et al. 2006), similar techniques have not been applied to model anticipated energy development and examine potential declines in species populations. Specific population decline estimates are needed to determine the need for special protection, as in the case of the Endangered Species Act (ESA) listing in the United States.

In this chapter, we use land use change buildout scenarios (wherein wells are placed on a simulated landscape) using a map of oil and gas development potential for portions of twelve states in the Intermountain West and published projections of reasonably foreseeable development from the Bureau of Land Management (BLM), as opposed to using expert or stakeholder input to create normative scenarios or visions (e.g., Nassauer and Corry 2004; Hulse et al. 2009). We then measure the impacts of the buildout scenarios on populations of greater sage-grouse (*Centrocercus urophasianus*; hereafter sage-grouse) across their eastern range (Stiver et al. 2006). Sage-grouse is a species for which impacts from energy development have been well documented (chap. 4), and spatially comprehensive and long-term data are available; it is a candidate species under the ESA. Here we provide a brief description of methods; full analysis and methods are reported in Copeland et al. (2009).

Forecasting Oil and Gas Potential

A contiguous and detailed map of oil and gas resource potential was needed to facilitate landscape-scale analysis because existing maps were limited in extent or too general. We created a probabilistic classification model of oil and gas resource potential to facilitate this type of landscape analysis. We generated a 1-square-kilometer prediction (map) using the nonparametric method Random Forests, developed to address statistical issues related to overfit and parameter sensitivity in Classification and Regression Trees (CART) models (De'ath and Fabricius 2000; Breiman 2001; Evans and Cushman 2009). The model used spatially referenced data on producing and nonproducing oil and gas wells with six predictor variables—geology, topography, bedrock depth, and anomalies in aeromagnetic gravity, isostatic gravity, and Bouguer gravity—variables used by geoscientists to predict where hydrocarbon deposits may occur (Ivanov 1985; Chen et al. 2000; Aydemir 2008). Bedrock geology maps show the age, distribution, and character of bedrock that lies immediately beneath the soils or surface. Our analysis used a national geology map available from the U.S. Geological Survey (Reed and Bush 2007). We used topographic data from the U.S. Geological Survey National Elevation Dataset (2007, www.usgs.gov) to indicate the location of fold and thrust belts where sedimentary rocks have been deformed by horizontal compression. Once compressed, tightly folded, and fractured, reservoir rocks may create pools for oil and gas to form (Newson 2001). The spatial distribution of the rock basement can be approximated using depth-to-bedrock data from the well database and indicates, at a coarse scale, where subsurface valleys and peaks of the basement rock are located. A 1-square-kilometer cell surface model of depth to bedrock was created from well depth information in the oil and gas wells database using inverse distance weighted interpolation (power = 1, number of points within radius = 12, maximum distance = 5,000 meters). Data on aeromagnetic and gravimetric anomalies depict spatial variations in subsurface rock density and magnetism and indicate features such as buried faults and the depth and location of the sedimentary rocks, both of which can be useful for hydrocarbon resource mapping (Kucks and Hill 2000). Data were used on the producing status of oil and gas wells within a 1-square-kilometer grid cell as the binary response variable with data acquired from the IHS Inc. database (2007, www.ihsenergy.com).

The study area was partitioned into coarse-scale geologic provinces (Fenneman and Johnson 1946), and modeling was conducted by province. Resulting model predictions were rescaled from 0 to 100, applied to each 1-square-kilometer grid cell, and mapped across the Intermountain West as oil and gas development potential, where 0 = low potential and 100 = high potential (fig. 10.1).

Accuracy (total number of correct classifications divided by the total number of sample points) varied in the individual models by 79.2–86.6 percent, with an overall accuracy of 82.9 percent. A full discussion of accuracy with accompanying tables is available in Copeland et al. (2009).

Despite sufficient model accuracy, the model uses coarse-scale data and thus provides a landscape- or regional-scale assessment of oil and gas potential. It is not intended to predict site-scale potential. Moreover, model predictions are constrained by the technology available at the time the wells were drilled and could be inaccurate if there are significant new advances in extraction technology.

Developing Buildout Scenarios

To predict and locate future oil and gas development, we ran two buildout scenarios—anticipated and unrestrained—by seeding the landscape with oil and gas wells according to the underlying development potential. The

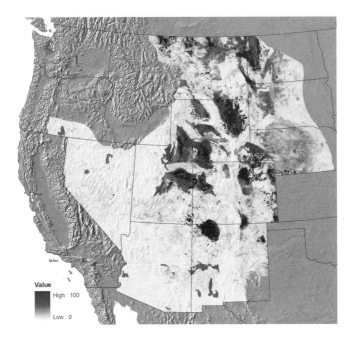

FIGURE 10.1. Oil and gas development potential in the U.S. Intermountain West (as modified from Copeland et al. 2009). This map shows the potential for oil and gas development from low to high (light to dark values). Black dots show producing (active or inactive) oil and gas well locations (well locations from IHS Inc.).

BLM is the federal agency responsible for managing mineral development on 114.5 million hectares (283 million acres) of public lands, including surface and subsurface mineral estate in the United States. The anticipated scenario was based on 20-year reasonable foreseeable development projections from the BLM's resource management plans. Where reasonable foreseeable development projections were unavailable, we estimated projections by doubling the number of wells permitted from 1996 through 2007 within a resource area. The unrestrained scenario allowed development in the highest quintile of oil and gas potential (model scores = 75–100). The BLM's estimates have been conservative: Colorado's White River Resource Area 1997 resource management plan predicted that fifty-six wells per year would be drilled, but the actual rate of drilling has been three times that since 2004 (U.S. Department of the Interior 1997). The number of current oil and gas leases across the study area also indicates that more lands are expected to be developed than resource management plans suggest. Using oil and gas leasing data from the BLM (U.S. Department of the Interior

2008), we calculated that 81 percent of federal lands with potential for oil and gas development (as defined in this scenario) have already been leased for oil and gas development. Therefore, we developed the unrestrained scenario to hedge against these uncertainties.

To place modeled oil and gas wells into the 1-square-kilometer cells available for development, we used Community Viz Scenario 360 Allocator Tool (Placeways LLC, Boulder, Colorado) to place wells into highest-probability cells first (using the map of oil and gas potential), then the next desirable, and so on, until all cells had met the specified demand at the specified density. If wells existed in a given cell, the model accounted for those wells in the demand calculation and added new wells until it fulfilled density limitations. The result—the number of wells expected in each cell—was written as an attribute to each 1-square-kilometer cell. We excluded lands where oil and gas development is currently prohibited, including national parks and national wilderness areas (National Atlas of the United States 2006) and no-surface-occupancy BLM lands. The resulting model placed a total of 95,867 wells at 16-hectare spacing (32 hectares in coal bed methane areas of the Powder River Basin, as per current regulations) in the anticipated scenario. The unrestrained scenario used the same constraints as the anticipated scenario but placed 260,953 wells in all areas with high oil and gas potential.

Assessing Impacts on Sage-Grouse Populations and Habitat

We used the two buildout scenarios to quantify the impacts of development on sage-grouse populations in their eastern range. Estimates of sage-grouse declines were restricted to areas in the eastern distribution of sage-grouse (fig. 10.2) (Connelly and Braun 1997; Stiver et al. 2006) because these populations are at greatest risk from energy development.

Oil and gas development is known to reduce sage-grouse populations at conventional well-spacing densities of 16–32 hectares (chap. 4). To determine whether sage-grouse would be affected in areas where development occurred, we relied on published studies by Doherty (2008) and Doherty et al. (2011) examining the relationship between oil and gas development and sage-grouse declines. These studies quantified the losses of both abundance and occurrence of sage-grouse populations due to oil and gas development across all leks in Wyoming, the largest segment of sage-grouse experiencing oil and gas development impacts in western North America. The average responses of leks in Wyoming to different

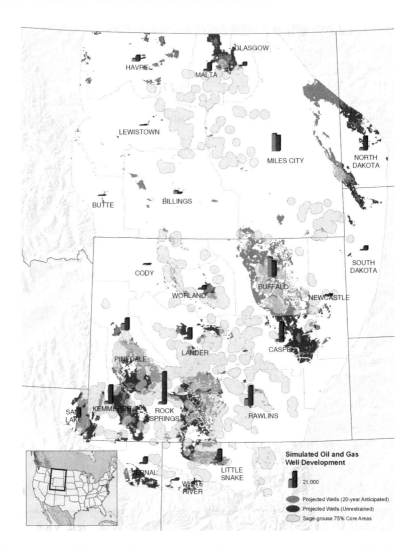

FIGURE 10.2. Oil and gas simulation results for the two scenarios (as modified from Copeland et al. 2009). This map illustrates the location and extent of expected development in the two scenarios. Areas in medium-gray shade depict growth for the anticipated scenario. Areas in dark gray depict growth for the unrestrained scenario. Bar graphs show the quantity of development projected for each scenario. Core areas (75% cores from fig. 4.3) for sage-grouse are shown to highlight expected areas of future conflict.

development intensities and amount of time in development compared with control populations experiencing no development were calculated for leks that were active in the last 11 years (Doherty 2008, tables 1 and 2, pp. 86–87). We applied the average responses from these tables to all leks throughout Management Zones I and II to predict future losses of sage-grouse to development. Current losses to energy development were discounted in calculations of predicted future losses of development by subtracting current losses at specific development intensities to anticipated losses at future development intensities.

The resulting model predicted a 7 percent population decline in the anticipated scenario and a 19 percent population decline in the unrestrained scenario, compared with 2007 lek population levels. These declines are in addition to the estimated rangewide population declines of 45–80 percent that have already occurred (Connelly and Braun 1997). The predictions for sage-grouse populations also imply impacts on other sagebrush-dependent species with known sensitivities to oil and gas development, such as pronghorn (*Antilocapra americana*), mule deer (*Odocoileus hemionus*) (Sawyer et al. 2006), Brewer's sparrow (*Spizella breweri*), sage sparrow (*Amphispiza belli*), and sage thrasher (*Oreoscoptes montanus*) (Ingelfinger and Anderson 2004). Projected impacts are distributed across sagebrush shrublands (3.7 million hectares) and grasslands (1.1 million hectares), with the remainder a mosaic of hayfields and irrigated croplands.

Implications from Buildout Scenarios

Without deviation from past development practices, we can expect a 7–19 percent population decline in sage-grouse. Furthermore, these impacts will be disproportionately borne by sagebrush and grassland ecosystems and the species that inhabit them (chap. 2). These results are based on the use of statistical models to forecast future change and the many assumptions inherent to this process. We based our buildout scenarios on projections from the most recent BLM planning documents available at the time and on the oil and gas potential model. The BLM estimates are frequently revised from new field discoveries and as technological advances influence resource extraction methods. Forecasted impacts to sage-grouse populations could be revised lower if directional drilling to reduce well pad density at the surface became more common (Sawyer et al. 2009). Our buildout scenarios are applicable across whole landscapes regardless of land tenure because we assumed that development could occur on any parcel of land, public or private, with the previously noted exceptions. Our estimates provide insights about the trajectory and eventual endpoint of oil and gas development, but the rate and exact location of development will be subject to additional factors not considered, such as market demand, the capacity to transport oil or gas to consumers, and federal air and water quality laws (e.g., Clean Air Act, climate change legislation).

There is no reason to believe that U.S. demand for fossil fuels will subside anytime soon. The increased pressure to meet these needs with domestic resources will probably amplify conflicts between conservation and development. For example, the systems that produce the world's food supply depend heavily on fossil fuels, especially liquid petroleum. Vast amounts of oil and gas are used as raw materials and energy in the manufacture of fertilizers and pesticides and as cheap and readily available energy at all stages of food production, from planting, irrigation, feeding, and harvesting through to processing, distribution, and packaging (Green 1978). Moreover, the land use efficiency of energy production will shape the continued reliance on hydrocarbons because this efficiency depends largely on the mode of production. McDonald et al. (2009) measured the areal impacts associated with various forms of energy development in area of affected land per terawatt hour of energy and found that efficiency varies over four orders of magnitude. Fossil fuels represent an efficient mode of energy production relative to other forms of energy development, particularly renewable energy (i.e., wind). This efficiency tradeoff is more dramatic when liquid fuels are considered. Biofuels are a potential low-carbon energy source that could be used to reduce carbon dioxide emissions relative to fossil fuels. However, biofuel alternatives to petroleum liquids fuels take the most space per unit power (McDonald et al. 2009), and whether biofuels offer carbon savings will depend on how they are produced (Fargione et al. 2008). A dramatic overhaul in energy usage patterns for transportation and agriculture must occur before shifts away from hydrocarbons will be possible.

Given the continued U.S. reliance on fossil fuels, the analysis presented here can inform planners and decision makers about where oil and gas development is anticipated and potential impacts to biodiversity and the tradeoffs between continued development and conservation needs. This analysis provides a general framework for using predictive models and buildout scenarios to anticipate impacts on species and the type of information needed for making decisions about special protections for species, such as ESA listing in the United States, and for biodiversity offsets (Doherty et al. 2011). The U.S. Fish and Wildlife Service, the agency that oversees ESA listing, faces difficult and complex decisions in determining whether current or future risk of species population declines warrants ESA protection. The economic ramifications of listing species are substantial, with estimated costs of recovery plans and their implementation reaching into the multimillions (Government Accountability Office 2006) if not billions of dollars for wide-ranging species such as sage-grouse. Prevention of listing through thoughtful consideration of threats and possible avoidance or mitigation strategies is likely to be less costly and more effective (Polasky 2008). This approach complements the proposed mitigation planning framework (chap. 9) by proactively identifying conservation and development conflicts and making mitigation recommendations consistent with sustainable development practices.

In the case of sage-grouse, 14–19 percent of the study area has high oil and gas development potential, but the development rights have not been sold; development in these areas could be avoided by removing these leases from auction or mandating other special protections by government management agencies. The sage-grouse core areas delineated by Doherty et al. (2011) provide information as to where state and federal agencies should focus conservation attention. The governor of Wyoming publicly supports the protection of sage-grouse core areas from new development, and other states are delineating core conservation areas and taking proactive steps to protect them. Areas already leased within sage-grouse core areas could be considered a priority for lease swaps, where government organizations, NGOs, and other private entities swap land or buy the initial lease back at cost from the purchasing company. Alternatively, companies could also be encouraged to forfeit their development rights with a perpetual no-surface-occupancy agreement, as part of the negotiation for enhanced access to exploration and development in other areas. Done in the right places, a creative combination of approaches could yield maximum benefit to species.

For many species experiencing population declines, multiple stressors are affecting their populations. The framework we present could be modified to consider not just one type of energy development, in this case oil and gas, but also wind, solar, coal, oil shale, and uranium, along with other stressors, such as residential development, invasive species, and pathogens. Because many of these stressors occupy disparate landscapes, this approach would account for cumulative impacts. For example, estimates suggest that meeting 20 percent of U.S. energy demand with wind would result in

50,000 square kilometers of land fragmented (U.S. Department of Energy 2008a), a significant portion of which would overlap with sage-grouse habitat. Moreover, wind development in the Intermountain West, in particular, does not correlate spatially with oil and gas development (chap. 2). Models and maps of multiple future threats are needed to fully quantify the overall future risk to sage-grouse.

Conclusion

The case of sage-grouse and oil and gas development in the Intermountain West is just one of numerous energy-versus-wildlife confrontations occurring across the globe, as population growth continues and previously undeveloped countries seek to westernize with modern appliances and transportation. Recognizing that we need energy to power our homes and businesses and that the Intermountain West holds tremendous potential energy resources, key decision makers need to quickly come to agreement about where these resources are best developed and best left alone. Modeling and envisioning future impacts can help justify proactive protection of places important to biodiversity, such as sage-grouse core areas, and underscore the ecological consequences of failing to do so. In Wyoming, the sage-grouse core area concept has provided a vision of where sage-grouse conservation is most needed, thus providing the foundation for more intelligent conversation between government and industry as new petroleum and wind leases are issued or renewed and new transmission lines permitted. Clearly, sage-grouse are not the only species likely to be affected by future development. Buildout scenarios evaluating impacts for other species using the techniques presented here need to be developed to gain a more complete picture of the anticipated impacts across a range of species, so decision makers can more fully evaluate the tradeoffs between development and wildlife.

Given rising global demands, all indicators point to an unprecedented amount of energy development in the future, with consequent biodiversity loss (International Energy Agency 2008). We hope to inspire regulators to use available technologies in mapping and modeling to forecast new impacts, and for policymakers to use this information to avoid business-as-usual development (Clark 2007; Ehrlich and Pringle 2008), in favor of proactive efforts to predict and avoid impacts in places crucial to the survival of a species. In the long run, this is likely to be the more ecologically robust, less costly, and more efficient course of action.

Acknowledgments

We thank the Montana and Wyoming State Offices of the BLM and Oil and Gas Commission for insights about oil and gas production on federal lands, Ronald Marrs for geological expertise in model design, Tom Rinkes and Tom Christiansen for initial thought-provoking discussion, Jody Daline for assistance compiling data on reasonable foreseeable development from BLM field offices, and Jeffrey Evans for statistical analysis support.

Chapter 11

Resource Policy, Adaptive Management, and Energy Development on Public Lands

MELINDA HARM BENSON

The challenge associated with integrating energy development and wildlife conservation on public lands is formidable. These pressures will continue, especially in the face of global climate change. Natural gas is viewed as an important bridge fuel to more renewable sources of energy (Roberts 2004). In addition to impacts associated with traditional oil and gas development, a host of new energy-related projects are emerging. Uranium mining is booming once again, and there are increasing pressures to develop energy transmission corridors, locate carbon capture and sequestration projects, and develop wind and solar facilities on public lands (U.S. Department of the Interior 2009a).

Fortunately, these challenges are accompanied by substantial opportunities to improve our ability to balance development needs with wildlife protection on public lands. Ecosystem-based adaptive management presents one such opportunity. In this chapter I examine the potential for adaptive management in the energy development arena, note some of the policy-level impediments to full implementation, and suggest some policy changes needed if the full power of adaptive management is to be realized. First, I examine the primary laws that govern leasing for energy development on U.S. federal lands and explore the potential use of both existing and proposed leasing provisions to adjust the pace and intensity of development to protect wildlife and other resources. Second, I discuss opportunities for more meaningful collaborative efforts on public lands in relation to the Federal Advisory Committee Act (FACA). The FACA was originally enacted to address issues of fairness among those seeking to influence the federal government but is now in need of reform to better support collaborative efforts. Finally, I address how federal laws governing the management of public lands can be used to incorporate adaptive management and make its application a legally enforceable requirement on public lands.

Adaptive Management and Energy Development on Public Lands

Adaptive resource management is an innovative environmental management strategy gaining influence with natural resource decision makers. It was developed to help incorporate the inevitability of scientific uncertainty into management actions involving natural systems (Holling 1978). A central tenet of adaptive management is that "management involves a continual learning process that cannot conveniently be separated into functions like 'research' and ongoing 'regulatory activities,' and probably never converges to a state of blissful equilibrium involving full knowledge and optimum productivity" (Walters 1986: 7; see also Walters and Holling 1990). Adaptive management represents a breakthrough in the complexity of our thinking about natural resource challenges. Rather than providing discrete conclusions based on science, adaptive management recognizes that our understanding of natural systems is constantly evolving. It reflects a willingness to test our assumptions about the natural environment to adapt and learn (Lee 1993). Federal agencies in the United States have recently begun incorporating adaptive management strategies. The U.S. Department of the Interior (DOI) and the U.S. Forest Service have issued administrative directives encouraging the use of adaptive management (U.S. Department of Agriculture 2007; Williams et al. 2009). The DOI officially adopted the National Research Council's definition of adaptive management:

> Adaptive management [is a decision process that] promotes flexible decision making that can be adjusted in the face of uncertainties as outcomes from management actions and other events become better understood. Careful monitoring of these outcomes both advances scientific understanding and helps adjust policies or operations as part of an iterative learning process. Adaptive management also recognizes the importance of natural variability in contributing to ecological resilience and productivity. It is not a "trial and error" process, but rather empha-

sizes learning while doing. Adaptive management does not represent an end in itself, but rather a means to more effective decisions and enhanced benefits. Its true measure is in how well it helps meet environmental, social and economic goals, increases scientific knowledge and reduces tension among stakeholders. (Williams et al. 2009: v)

Several key steps for implementing adaptive management are outlined in the DOI approach. The first is to ensure stakeholder commitment to adaptively manage the enterprise for its duration, which requires that interested parties be involved early in the process. The second is to identify clear, measurable, and agreed-upon management objectives to guide decision making and evaluate management effectiveness over time. Then, sets of potential management actions and models that characterize different ideas about how the system works are identified. Next, a monitoring plan is developed and implemented to track resource status. Management actions are chosen based on objectives, and monitoring is used to track system responses to management actions by comparing predicted and observed changes in resource status. Finally, monitoring information gathered to improve the agency's understanding of resource dynamics is integrated back into the decision-making process.

The next few years will be especially critical in the implementation of adaptive management as agencies struggle to implement strategies within current legal and regulatory structures. The remainder of this chapter focuses on ways to work within those structures to make adaptive management a valuable and enforceable tool for protecting wildlife and other resources.

The Mineral Leasing Act and the Capacity to Adapt

Leasing of fluid minerals on federal lands takes place under the Mineral Leasing Act of 1920 (MLA), as modified by the Federal Onshore Oil and Gas Leasing Reform Act of 1987 (30 USC §§ 181 et seq.; 10 USC §§ 226 et seq., respectively). As defined by the statutes, fluid minerals include oil and natural gas, including coal bed methane. In contrast, hardrock minerals on federal lands including uranium are governed by the 1872 Mining Law. The MLA is implemented by two DOI agencies. The Bureau of Land Management (BLM) oversees the issuance of leases and actual development for oil and gas, and the Mineral Management Service collects, accounts for, and distributes revenues associated with mineral production. Leases are issued through a bidding process, generally with a primary term of 10 years. Once a parcel is leased, reasonable diligence must be exercised by the leaseholder to develop the resource. The primary term lease provisions create an incentive to proceed quickly with development to move into the production phase that will both secure the leaseholder's interest and provide an economic return on the investment.

An oil and gas lease is essentially a contract, defining the rights and responsibilities of both parties. A common perception—one often encouraged by the BLM—is that the government has little influence on the pace or nature of development once an area is leased. But the agency actually retains a great deal of influence over how development occurs, through both standard leasing provisions and additional stipulations that may be attached to a lease. Three main sections under standard leasing provisions, common to almost all leases, reserve agency authority to address environmental concerns (as per Offer to Lease for Oil and Gas Form 3100-1, June 1988). First, section 4 specifically "reserves the right to specify the rates of development and production in the public interest" and has been recognized by Interior Board of Land Appeals as having "vest[ed] [BLM] with adequate authority to protect wildlife values" (*National Wildlife Federation*, 169 Interior Board of Land Appeals 146, 164 [2006]; see also *Wyoming Outdoor Council*, 147 Interior Board of Land Appeals 105 [1998], and *Powder River Basin Resource Council*, 120 Interior Board of Land Appeals 47 [1991]). Next, section 6, titled "Conduct of Operations," also addresses the need to minimize environmental impacts, stating, "Prior to disturbing the surface of the leased lands, the lessee shall contact the lessor to be apprised of procedures to be followed and modifications or reclamation measures that may be necessary. Areas to be disturbed may require inventories or special studies to determine the extent of impacts on other resources." This provision anticipates that additional environmental protection measures may be required after leasing but before actual development, and it reserves the right of the BLM to make modifications as necessary to address environmental concerns. With regard to wildlife specifically, it notes that "if, in the conduct of operations, lessee observes or encounters any threatened or endangered species . . . or other specific resources that are statutorily protected, or substantially different, unanticipated, environmental effects, lessee shall immediately contact lessor. Lessee shall cease any operations that would result in the destruction of such." Finally, section 7, titled "Damages to Property," states, "If impacts are substantially different or greater than normal . . . lessor reserves the right to deny approval of operations." This provision also recognizes that unanticipated impacts may be associated with development and reserves the right to restrict development. Each of these sections of the standard lease is a part of the contractual agreement between the government and the leaseholder seeking to develop mineral resources on public lands.

In addition to these standard leasing provisions, leases are often supplemented with additional stipulations. Stipulations address specific concerns of a particular area, and wildlife protection is often the goal. For example, the BLM can place seasonal stipulations that limit activity during certain times when wildlife are particularly vulnerable, such as restricting drilling activity during the winter in crucial winter range for mule deer (*Odocoileus hemionus*) or limiting surface disturbance in the early spring where there are known greater sage-grouse (*Centrocercus urophasianus*) leks.

Unfortunately, the BLM seldom exercises its contractual authority under the standard lease provisions. Even when it does, it often backs down in the face of industry pressure. Wildlife stipulations, where they exist, are often waived. The BLM's Pinedale Resource Area in southwest Wyoming granted 97 out of 113 requests from oil and gas operators to waive stipulations designed to protect raptors, 103 of 116 requests to waive wildlife winter range protections, and 119 of 170 requests to waive stipulations to protect sage-grouse (The Wilderness Society 2005).

The BLM's equivocation on protecting wildlife in the face of energy development reflects the challenge associated with balancing its land management responsibilities and its role as a facilitator of energy development under the MLA. While reserving authority to address environmental impacts, the MLA includes no explicit requirement to balance energy development with other land uses. When the MLA was amended by the Federal Onshore Oil and Gas Leasing Reform Act in 1987, the main focus was to address inefficiencies in the leasing process, make leasing more competitive, and boost royalty payments (Sansonetti and Murray 1990). At the time, there was debate regarding whether to include provisions addressing the impacts of oil and gas development to other lands uses. Those amendments did not make it into final legislation, but as a compromise, Congress directed the National Academy of Sciences (NAS) to conduct a study on the manner in which oil and gas resources are considered in land use plans. The NAS did the study and provided Congress with several recommendations (NAS Committee on Onshore Oil and Gas Leasing 1989).

The first recommendation focused specifically on areas that had high potential for both energy development and land use conflicts. The NAS recommended that these areas be leased with stipulations that allowed only for exploratory wells, allowing information gained during exploration to guide later decisions regarding whether to allow further development. The NAS recommended that leaseholders be reimbursed for the costs of the lease and exploration if it turned out that other land use values precluded mineral development. The second recommendation, which the NAS recommended apply to all leases, was to include a stipulation that "preserves the government's flexibility to control and if necessary, to prohibit, activities on leases that pose serious and unacceptable impacts on other values, but with the provision that a lessee would be reimbursed for its direct costs in acquiring and developing its lease if further exploration and development is prohibited" (NAS Committee on Onshore Oil and Gas Leasing 1989: 127). Both of these recommendations, if implemented, would provide the BLM with further tools for engaging in ecosystem-based, adaptive management.

Another alternative would be to simply attach adaptive management stipulations to leases. Such stipulations would place leaseholders on notice that the pace and intensity of development would be tied to experimentation and learning, as necessary, to protect other resource values. Oil and gas development on federal lands will continue. For this development to proceed in a manner compatible with an adaptive management approach, the BLM must be more willing to exercise its existing authority and perhaps create some new authority to achieve a more balanced approach to energy development. These changes have the potential to make adaptive management more attractive to industry. If leaseholders are allowed to recoup their costs if they are prohibited from developing their leases, the need for flexibility can be balanced by providing some certainty.

New renewable energy projects, such as wind and solar, currently receive little of the potential environmental protection available under formal leasing processes like those provided under the MLA. Instead, the BLM uses general authority to grant rights of way, traditionally used for projects such as transmission lines, for locating these projects (42 USC § 1761(a)(4); 43 C.F.R. 2800). This, combined with pressures to expedite renewable energy projects, threatens to lead to the same or worse mistakes as those made in conventional energy development (BLM 2005; DOI 2009b). Just as with traditional energy development, new resources should be paced with learning to address impacts on wildlife and other resources.

Collaborative Engagement on Public Lands

One of the most undervalued yet critical aspects of adaptive management is effective collaboration between stakeholders (Margoluis et al. 1998). Public lands have from their inception been fraught with conflict and competing interests, creating "wicked problems" characterized by radical uncertainty and social and ecological complexity (Rittel and Webber 1973). Although it is time-consuming, working with multiple stakeholders increases the likelihood of a project's success (e.g., see the Collaborative Adaptive Management Network at www.adaptivemanagement.net). Consequently, most models of adaptive management outline a first stage that includes the development of a shared conceptual model among stakeholders. The DOI's new technical guidance for adaptive management specifically emphasizes that ensuring stakeholder commitment to adaptive management for the duration of the process is a key first step.

Unfortunately, meaningful collaboration on public lands is hampered by the FACA, which was enacted in 1972 in response to increasing concern over escalating influence by industry groups and others on government decision making. The statute broadly defines the term *advisory committee* to include "any group, with one or more public members, created by law or established or 'utilized' by an agency or the President" (5 USC app. 2 §§ 1–15). Collaborative processes typically fall squarely within this definition. Once triggered, the FACA requires an onerous process for the establishment and operation as an official FACA committee. The result is a chilling effect in which collaborative processes are often avoided or are unsuccessful because they cannot meet statutory requirements (Beierle and Long 1999). Thus, the FACA is a primary legal barrier to ecosystem management, and agencies often decline to obtain outside input rather than face future litigation over compliance with the FACA (Government Accountability Office 1998). Even if agencies are willing to put forth the effort to go through the required administrative process, it often takes years to receive a formal FACA charter. This delay severely limits citizen advisory groups in many actions involving natural resources, which often are time-sensitive. The FACA also requires committees to receive new charters every 2 years, a particularly cumbersome constraint in natural resource contexts, where decisions are often implemented over a much longer period. There is also a limit to the number of committees that can exist at any given time, further limiting opportunities for collaboration.

FACA reform is needed to better use collaborative processes necessary for adaptive management implementation. The FACA should be made more flexible and should encourage, rather than hinder, adaptive management. Beierle and Long (1999) make several recommendations for reform, two of which are directly relevant in this context. First, they recommend lifting the administrative ceiling on the number of FACA committees that can exist at any given time. This would allow agencies to engage in collaborative processes more often and benefit from public involvement. Second, they recommend streamlining the procedural requirements for establishing an advisory committee. Current procedures result in unreasonable delays that undermine collaborative process and its effectiveness. Another more sweeping suggestion for reform is to exempt certain categories of committees from the FACA altogether. Collaborative processes designed to implement adaptive management may be appropriate for such an exemption.

Making Adaptive Management the New Paradigm on Public Lands

The final key step on the path forward is the reorientation of management efforts on public lands to make them more amenable to adaptive strategies. Although a comprehensive examination of public land policy is beyond the scope of this chapter, here I focus on current efforts to implement adaptive management under two main statutes governing public land management: the National Forest Management Act and the Federal Land Policy and Management Act. I also consider the National Environmental Policy Act (NEPA), which applies to all major federal actions on public lands, as a potential vehicle for implementing adaptive management. Finally, the U.S. Endangered Species Act is discussed as the current "emergency room" approach to wildlife protection. Without substantial land management reform, this will continue to be the dominant strategy for addressing the impacts of energy development on wildlife.

Current Efforts by Land Management Agencies

The Federal Land Policy and Management Act and the National Forest Management Act are the primary laws governing the management of federal lands generally available for energy development. The National Forest Management Act governs 78.1 million hectares (193 million acres) of forests and grasslands managed by the U.S. Forest Service. The Federal Land Policy and Management Act governs the DOI's 103.6 million hectares (256 million acres) of BLM lands. The BLM also manages the federal government's 283.3 million hectares (700 million acres) of subsurface mineral estate, although land-based managers such as the Forest Service have a role in determining where leasing occurs on the land they manage. Both national forests and BLM lands are governed by multiple-use, sustained-yield principles requiring them to balance a number of competing interests on public lands, including recreation, grazing, timber harvesting, and mineral development, in addition to watershed, wildlife, scenic, scientific, and historical values (43 USC §§ 1701 et al.; 6 USC §§1600 et al., respectively).

These management mandates are sufficiently vague to encompass adaptive management approaches but do not provide the regulatory home needed for adaptive management. "The disconnect between adaptive management *in practice* and adaptive management *in law* is quite palpable. Today's practitioner of natural resource law is bombarded with adaptive management. It is firmly entrenched in natural resource agency practice from headquarters to field level. It shows up in land management plans, resource development permits, and agency guidance documents. Yet it appears almost nowhere as codified statutory or regulatory text, and it is dealt with significantly by only a handful of judicial opinions. . . . No other principle of natural resources law has so deeply permeated the practice on the basis of so little mention in law" (Ruhl 2008: 11-3).

Both the U.S. Forest Service and the BLM have used mainly departmental orders, instructional memoranda, and agency manuals to integrate adaptive management principles. In March 2007, all bureaus within the DOI were ordered to use adaptive management whenever possible, and this order was incorporated into the DOI *Departmental Manual* in February 2008 (i.e., Secretary of the Interior Secretariat Order 3270 and BLM Adaptive Management Implementation Policy). The Forest Service is implementing a process called environmental management systems (EMS) on all administrative units and incorporates adaptive management as a fundamental principle (U.S. Department of Agriculture 2007).

Although these departmental directives are an important beginning, they are insufficient. Without more specific statutory or regulatory grounding, commitments to adaptive management are generally not binding on the agency (Fischman 2007). This means that those outside the agency are not able to enforce the agencies' commitment to adaptive management. Enforceability is important because, historically, judicial interpretation has been necessary to establish the details and define the duties and expectations of agency mandates and ensure implementation (Nie 2008). Without more specific and enforceable legal grounding, adaptive management principles are in danger of losing their legitimacy, as "agency speak" with little meaning. Already, critics of adaptive management view it as an excuse to allow agencies an unreasonable amount of discretion (Doremus 2001; Houck 2009a).

NEPA as a Potential Regulatory Home for Adaptive Management

NEPA may provide a regulatory home for adaptive management. NEPA applies to all major federal actions that have a significant impact on the human environment. NEPA is often referred to as having twin aims in providing better-informed agency decisions and involving the public in important natural resource decision making (42 USC § 4331 et al.). NEPA section 102 requires agencies to provide a detailed statement (known as an Environmental Impact Statement) outlining the environmental impacts of the proposed action, listing alternatives to the action, and explaining whether the action necessitates any irretrievable or irreversible commitments of resources by the agency. It also establishes a new agency, the Council on Environmental Quality, to help guide NEPA implementation.

The original drafters of NEPA had something more in mind, however. Section 101 of NEPA declared that it was "the continuing policy of the Federal Government . . . to use all practicable means and measures, including financial and technical assistance, in a manner calculated to foster and promote the general welfare, to create and maintain conditions under which man and nature can exist in productive harmony, and fulfill the social, economic and other requirements of present and future generations of Americans" (42 USC § 4331 et al.). This provision, sometimes called the substantive provision of NEPA, was intended to require agencies to make not just well-informed but also environmentally sound decisions balancing environmental and economic concerns (Caldwell 2002).

This substantive goal never came to fruition because it was quickly undermined by court interpretations of the law, which held that NEPA's section 101 was an aspirational statement lacking the necessary detail for enforcement. In its early interpretations of the statute, the U.S. Supreme Court held that NEPA, while establishing "significant substantive goals for the Nation, imposes upon agencies duties that are essentially procedural. . . . NEPA was designed to insure a fully informed and well-considered decision" (*Stryker's Bay Neighborhood Council Incorporated v. Karlen*, 444 U.S. 23 [1980], quoting *Vermont Yankee Nuclear Power Corporation v. Natural Resources Defense Council Incorporated*, 435 U.S. 519 [1978]). As a result, NEPA essentially became a series of steps requiring federal agencies to take a hard look at the potential environmental consequences of their action, but not requiring them to take any specific action to protect the environment or balance competing concerns.

Adaptive management may provide a way to reclaim the potential of NEPA. Because NEPA applies to all federal agencies, integration of adaptive management principles into NEPA's requirements would provide important depth and consistency to both agency and court interpretations of adaptive management methods and requirements. In many ways, the underlying goals and principles of NEPA and adaptive management are inherently compatible, and efforts are being made to integrate adaptive management principles into NEPA processes (Dreher 2005). In 2002, the Council on Environmental Quality created a NEPA Task Force to address issues associated with modernizing NEPA, including the need to better integrate adaptive management (National Environmental Policy Task Force 2003). A year later, the task force did come forward with a series of recommendations, which included the recommendation that the Council on Environmental Quality convene an Adaptive Management Working Group that would assist in promulgating regulations specific to adaptive management. This has not yet happened.

In its current form, NEPA has been criticized for actually hampering adaptive management (Benson 2009). Like most statutes of its era, NEPA did not address the iterative decision-making process necessitated by adaptive management. Its requirement of a detailed statement outlining the anticipated impacts of a proposed agency action reflects what has been called a front-end approach that assumes all the information needed to make a decision is already available (Ruhl 2005; Thrower 2006). It also assumes a single action rather than ongoing activities that must be monitored and adjusted as necessary to achieve management goals.

Efforts are being made to integrate adaptive management principles into NEPA processes, but more fundamental reform of NEPA is needed to make adaptive management an actual requirement. Within the DOI, bureaus have been directed to "use adaptive management, as appropriate, particularly in circumstances where long-term impacts may be uncertain and future monitoring will be needed to make adjustments in subsequent implementation decisions" (43 C.F.R. § 46.145). Proponents of NEPA reform usually argue for new amendments to the law, but reconfiguration of NEPA could be achieved without new legislation (Haugrud 2009). The Council on Environmental Quality, in its role as the main interpreter of NEPA, could implement new regulations to require adaptive management for agency actions that trigger NEPA and also outline specific protocols and methods for implementing adaptive management. Reform of NEPA would admittedly be a significant statutory reinterpretation, but it would be permissible under general principles of administrative law because, as noted earlier, NEPA's original intent was to play a more substantive role in agency decision making. This issue was recently addressed by the U.S.

Supreme Court, which held that federal agencies can adopt new regulations that reinterpret a statutory provision previously defined by the courts under certain circumstances (*National Cable and Telecommunication Association v. Brand X Internet Service* [Brand X], 545 U.S. 967 [2005]). Those circumstances include an interpretation that the court does not hold as the only reasonable interpretation of the statute. In the case of NEPA, none of the early interpretations of its requirements were based on a clear reading of the statutory language. On the contrary, the courts ignored the substantive statutory language in section 101 and instead focused on the more procedural requirements outlined in section 102. For this reason, NEPA can be reinvigorated without statutory amendment.

Reclaiming NEPA's potential has been advocated by others, particularly those seeking more collaborative approaches to environmental decision making. In 2000, Congress directed the U.S. Institute for Environmental Conflict Resolution to charter a FACA committee to investigate "strategies for using collaboration, consensus building and dispute resolution to achieve the substantive goals" of NEPA (U.S. Institute for Environmental Conflict Resolution 2005: 3). The National Environmental Conflict Resolution Advisory Committee was formed and issued its findings in 2005 (U.S. Institute for Environmental Conflict Resolution 2005). The report promoted resurrecting NEPA's section 101 to advance federal agency use of collaboration and environmental conflict resolution. Among the recommendations was an acknowledgment of the role of adaptive management. In recognizing adaptive management as the future of natural resource management, the National Environmental Conflict Resolution Advisory Committee recommended that the United States "identify ways to expand its leadership in developing applications of collaborative monitoring in the context of alternative dispute resolution and adaptive management" (U.S. Institute for Environmental Conflict Resolution 2005: 20).

Several specific reforms are necessary if NEPA is to be reconfigured to integrate adaptive management. First, and foremost, NEPA's front-end approach must be reworked to reflect the iterative processes necessitated by adaptive management. As a practical matter, some agencies are already taking steps in this direction and are increasingly using a process called tiering. Tiering allows agencies to sequence NEPA analysis starting with a broad programmatic Environmental Impact Statement that addresses larger policy issues, or the initial stages of a program are then supplemented with more site-specific analyses (40 C.F.R. § 1508.28). The current regulatory guidance specifically encourages tiering in situations in which decision making occurs in stages (40 C.F.R. § 1502.20).

The regulations also anticipate situations in which information is incomplete or unavailable. In such cases, regulations direct the agency to provide a "statement that such information is incomplete or unavailable" and include a "statement of the relevance of the incomplete or unavailable information to evaluating reasonably foreseeable significant adverse impacts on the human environment," along with a summary of the "credible scientific evidence which is relevant" (40 C.F.R. § 1502.22). Adaptive management would incorporate both approaches, building the informational needs into the conceptual model and designing management approaches that begin to fill in the gaps and guide future action.

As can be seen in this brief overview, current regulations provide some basis for adaptive management, but full integration of adaptive management's iterative approach would require developing specific protocols to guide implementation, including when new information triggers the requirement for a supplemental Environmental Impact Statement. Without real reform, programmatic tiering can become a shell game in which various NEPA documents cross-reference each other but fail to provide a comprehensive assessment of the environmental impacts of a project (Benson 2009).

The second major reform needed to reconfigure NEPA is to require effective monitoring, an essential component of adaptive management that is designed to assess management experiments and provide the feedback necessary for further knowledge integration and iteration. DOI guidance specifically emphasizes that "monitoring is used in adaptive management to track system behavior, and in particular to track the responses to management through time. In the context of adaptive management, monitoring is seen as an ongoing activity, producing data after each management intervention to evaluate the intervention, update the measures of model confidence and prioritize management options in the next time period" (Williams et al. 2009: 33).

Currently, NEPA regulations state that agencies "may provide for monitoring to assure that their decisions are carried out and should do so in important cases." It falls short of actually requiring monitoring (with one exception noted later), and monitoring, as a practical matter, is often abandoned when budget constraints require agencies to cut back (Benson 2009). NEPA has the potential to provide the legislative mandate for monitoring, which would allow agencies to build the necessary funding into their budgets and, when appropriate, require financial support from those seeking to develop resources on public lands (DeLuca et al. 2008).

The third key element of NEPA reform needed to achieve adaptive management is an affirmative obligation to engage in mitigation of environmental impacts. Mitigation is currently defined by the NEPA regulations to include avoiding the impact altogether by not taking a certain action or parts of an action; minimizing impacts by limiting the degree or magnitude of the action and its implementation; rectifying the impact by repairing, rehabilitating, or restoring the affected environment; reducing or eliminating the impact over time through preservation and maintenance operations during the life of the action; and compensating for the impact by replacing or providing substitute resources or environments (40 C.F.R. §1508.20). Although NEPA currently requires an examination of "appropriate mitigation measures not already included in the proposed action or alternatives," it does not generally require their implementation (40 C.F.R. § 1502.14[f]). To the extent to which mitigation has been encouraged under NEPA, it has been done indirectly by agencies seeking to avoid NEPA requirements by developing mitigated "findings of no significant impact" that allow them to avoid the requirements of an Environmental Impact Statement (Haugrud 2009). Where the NEPA process does end up choosing an alternative requiring mitigation, NEPA requires the agency to also engage in monitoring (the exception noted earlier) and allows the agency to condition its approval of permits, funding, and other activities on the mitigation required by the decision (40 C.F.R. § 1505.2 and 1505.3). NEPA falls short of consistently requiring mitigation, however, which is a key ingredient to an adaptive management approach (Karkkainen 2002). A revitalized NEPA would combine monitoring and mitigation to achieve the substantive goals that were NEPA's original intent.

The final reform necessary is to insert NEPA back into agency planning. Just as early cases eviscerated NEPA section 101, court decisions also narrowly defined "agency action," exempting many planning processes (Houck 2009b; Mandelker 2009). This has long been regarded as one of the major failings of NEPA implementation, leaving evaluations of cumulative impacts as individual project-level decisions and therefore missing the opportunity whether to develop at all at the planning stage. Adaptive management requires agencies to use planning as an opportunity to develop a conceptual model and management objectives to guide agency decisions. NEPA at the planning stage, as integrated toward a more iterative process, is a common-sense reform necessary to meet NEPA's original intent.

All these reforms—integrating an iterative process, requiring monitoring and mitigation, and reinstating NEPA's hard look at the planning stage—have the potential to incorporate adaptive management as the new management paradigm on public lands. If implemented, they would provide agencies with the necessary flexibility to address impacts on wildlife and other natural resources before habitat degradation and other threats to species necessitate the more drastic approaches discussed below.

The Endangered Species Act: The Emergency Room Approach

The Endangered Species Act (ESA) will be of increasing importance, particularly in the absence the reforms mentioned earlier. All the reforms suggested in this chapter reflect an attempt to address challenges associated with balancing energy development and wildlife protection before significant problems occur and decision making becomes crisis-driven. By contrast, the ESA is an emergency room approach to species conservation, providing protection only for species that are listed as threatened with or in danger of extinction. Once a species is listed, the ESA provides an unparalleled level of legal protection against further decline. Though not exclusively applicable to federal lands, the ESA nonetheless greatly influences public land management decisions. Among its key provisions is a requirement that federal land managers ensure that their actions do not jeopardize the species or its critical habitat. This process, which involves consultation with the U.S. Fish and Wildlife Service, can be both time consuming and contentious. It limits the capacity of agencies to engage in the types of experimentation required by adaptive management. It can also result in severe limits on extractive activities, as was the case with efforts to protect the now-famous northern spotted owl (*Strix occidentalis caurina*) from habitat destruction resulting from logging in the Pacific Northwest.

In the absence of a more comprehensive requirement to protect biodiversity in the United States, the ESA's strong tools for conservation will continue to be the major driver in many controversies on public lands. Even the threat of a species being placed on the ESA threatened or endangered list can encourage federal land managers and others to implement measures to protect the species and thereby avoid listing. This can occur formally in the form of Candidate Conservation Agreements or informally through cooperative interagency efforts to protect habitat.

In the case of energy development, the candidate listing of the greater sage-grouse is a major focus (U.S. Fish and Wildlife Service 2008). As discussed in chapter 4, Wyoming and other states are developing conservation plans to identify "strategies and commitments for the purpose of improving sage-grouse numbers and precluding the need for listing under the Endangered Species Act" (Northeast Sage-Grouse Working Group 2006). Wyoming has established core habitat areas for sage-grouse protection (State of Wyoming Executive Order 2008-2). The BLM has also developed a

National Sage-Grouse Habitat Conservation Strategy. However, the alleged failure of the BLM to implement that strategy into its land use planning process is now the subject of a major legal challenge (*Western Watersheds Project v. Kempthorne* 2008 Amended Complaint Number 08-cv-516-BLW).

Conclusion

Just like an emergency room, the ESA is a necessary safety net, but it is insufficient for overall ecosystem health. Efforts to address the needs of one species, or even several species, do not address the complexities and needs of an entire ecosystem. A more sustained and comprehensive approach is needed, and adaptive management provides an opportunity to implement such an approach. For adaptive management to be used successfully, the path forward will require agencies to pace energy development to account for uncertainties about the impacts on wildlife and other resources, and to incorporate learning into their management efforts. This can be achieved through more aggressive enforcement of current leasing provisions that protect natural resources, and perhaps through the explicit incorporation of adaptive management strategies via new stipulations.

The path forward will also require more effective collaborative efforts involving the many stakeholders involved in developing energy on public lands. Current efforts tend to be focused on interagency cooperation, but effective adaptive management requires broader participation from industry groups, conservationists, local communities, and others. FACA, the federal law governing citizen participation in agency decision making, must be reformed to allow more stakeholder involvement in collaborative efforts.

However, real reform will require the integration of adaptive management as the new paradigm on public lands. Although some initial efforts are under way, successful implementation of adaptive management will require making its procedures and protocols legally mandated and enforceable. NEPA, the law that once held great promise for achieving environmental protection on federal lands, has the potential to provide the much-needed regulatory home for adaptive management. These reforms can be achieved without statutory amendment, and because NEPA applies to all federal agencies, it has the potential to provide the consistency and depth needed for effective implementation. Pressures to develop energy resources on public lands continue, and meaningful efforts to protect wildlife will require us to take this new path forward.

Chapter 12

Community-Based Landscape Conservation: A Roadmap for the Future

Gregory A. Neudecker, Alison L. Duvall, and James W. Stutzman

Energy independence, increasing domestic demand, and resulting impacts of development on wildlife portend a critical challenge to conservation. Energy development is not new to the West, but the accelerating rate of energy development and the expanding overall human footprint on western landscapes leave conservationists scrambling for ways to reduce anthropogenic impacts on wildlife. In a perfect world, there would be enough time and money to save everything, but instead we face a growing list of imperiled species, declining extent and condition of habitats, elevated risk to intact and functioning landscapes, and ever-present budget shortfalls. These conditions and constraints demand an overall approach based on conservation triage, defined here as the prioritization of landscapes to which limited resources are allocated to maximize biological return on investment (Bottrill et al. 2008, 2009). Once seen as a defeatist conservation ethic, triage is now viewed as a crucial approach to maintaining biological resources, in contrast to providing palliative care to already degraded systems (Schneider et al. 2010). Indeed, the conservation paradigm has shifted in scale and practice from small and reactive to large and proactive approaches to implementing landscape conservation before the opportunity to do so is lost.

Landscape planning has typically been a biological endeavor, but the real key to implementing lasting conservation is in working with people to maintain rural ways of life that are compatible with biological goals.

Community-based conservation originated in the 1980s in response to criticism of major international organizations for designing and implementing conservation with little input from local communities (Chambers 2007). In the international conservation arena, local communities voiced disapproval of large-scale, capital-intensive, and centrally planned projects that excluded them from participation and benefit (Kumar 2005). The rise of community-based conservation also resonated in the United States as agencies explored a related but somewhat independent trend away from top-down, regulatory-based, and expert-driven resource management toward voluntary, incentive-based conservation with broad public and community inclusion in land management programs (Weber 2000; Wondolleck and Yaffee 2000). Today, community-based conservation has evolved from a theoretical argument against actions that exclude humans to integrated approaches that embrace equally the societal and biological aspects of conservation (Horwich and Lyon 2007). Cornerstone principles include local participation, sustainable natural and human communities, inclusion of disempowered voices, and voluntary consent and compliance, rather than enforcement by legal and regulatory coercion. Win–win outcomes are sought, with all stakeholders at the table (Weber 2000; Nie 2003).

Our goal in this chapter is to provide readers with a vision of the next frontier in conservation: linking landscape conservation with people and communities. We suggest a "3P" formula for success that includes selecting the right places, people, and projects to manage collectively the multiple stressors affecting priority landscapes and maximize conservation investments. We conclude by highlighting the relevance of community-based conservation to energy development. The chapter is intended to provide a roadmap that empowers decision makers with strategies for maintaining biological values in the face of expanding human footprints, increases capacity for local and regional conservation initiatives, and supports local practitioners with ways to facilitate conservation while acting as strong advocates for people and communities.

Strategic Landscape Conservation: Working in the Right Places

Conservationists around the country grapple with the best ways to deliver effective programs to benefit species, ecosystems, and the human communities within them, which is a daunting task given the diversity of perspectives, experiences, philosophies, and mandates involved. The science, and indeed the art, of first identifying and subsequently delivering conservation in priority landscapes continues to gain support as a prevailing paradigm. Still, many programs are best characterized as opportunistic conservation that takes a shotgun approach to deciding where to work and gauges success by the total amount of habitat treated or manipulated. From a practitioner's standpoint, the projects may have sizable benefits at the ownership or habitat scale but may still be isolated in the midst of fragmented landscapes. Others are using individual fish and wildlife species or a suite of species as the justification to prioritize conservation actions. Although in theory the latter approach has merit, it generally does not account for broader ecosystem or human dimensions. In our experience, delivering conservation in the right place is best practiced by integrating the two ideologies by focusing on whole landscapes and on a suite of focal species that reside within those systems.

We suggest a proven ten-step Conservation Focus Area (CFA) approach, developed by the U.S. Fish and Wildlife Service's Montana Partners for Fish and Wildlife Program (herein "Partners Program"), as an advanced model for prioritizing where to work, especially given limitations in funding and capacity for implementation. These ten steps provide filters, moving from broad to narrow criteria for prioritization of areas to invest resources and time (table 12.1).

In application, the Partners Program model of biological planning begins by locating CFAs within geographic areas or ecoregions. Ecoregions

TABLE 12.1. A 10-step Conservation Focus Area (CFA) approach.

1	Use geographic areas or ecoregions as a foundation for planning.
2	Select a representative set of focal species.
3	Initiate biological planning by compiling in a geographic information system all relevant habitat and population data for focal species.
4	Identify initial overlap in conservation plans between state, federal, and non-government partners.
5	Consult with partners to view strengths and weaknesses in biological data.
6	Draft initial set of CFAs.
7	Use landscape intactness and public–private ownership patterns to compare draft CFAs.
8	Assess existing community-based conservation groups already working in identified CFAs.
9	Evaluate realized and potential future threats to CFAs.
10	Formally select final set of CFAs.

cover large areas that contain geographically and climatically distinct assemblages of natural communities and fish and wildlife species, in contrast to jurisdictional lines for management such as state or county boundaries (Bailey 1995). The United States has sixty-four distinct ecoregions, seventeen of which are located in the Rocky Mountain West.

The next step involves selection of focal species for each distinct geographic area. Focal species can help provide a practical bridge between single- and multiple-species approaches to wildlife conservation and management (Mills 2007). However, more than 1.5 million species can present major challenges to implementing conservation actions in a way that is logistically, financially, and politically feasible. One viable solution is to develop inference about the larger community or ecosystem based on a subset of species in the system. Four categories of focal species—flagship, umbrella, indicator, and keystone species—can be used collectively to help expand beyond single-species management and toward ecosystem management (box 12.1).

Once ecoregions and focal species have been selected, biological planning and conservation design follow. Biological planning is the systematic application of scientific knowledge about species and habitat conservation (Johnson et al. 2009c). Planning includes articulating measurable population objectives for selected focal species, identifying what may be limiting populations below desired levels, and compiling models that describe how populations are expected to respond to specific conservation actions. Conservation design is a rigorous mapping process, based on geographic information systems, that predicts patterns in the ecosystem and develops species-specific models, associated habitat objectives, and maps of biodiversity and species richness. The maps are produced by applying empirical models to spatial data (Johnson et al. 2009c).

The next steps in the CFA approach include consultation with in-house staff, state and federal agencies, and private conservation partners. First, the data are compared with other biologically based public and private plans that cover the same geographic areas, such as state wildlife agency plans. This step is followed by involving partners in the process of viewing the ecoregional maps, with interpretation and discussion of possible gaps in knowledge before selection of the CFAs.

At this stage, conservation practitioners have sufficient data to draft a set of potential CFAs. For example, the Partners Program and its partners identified eighteen potential CFAs across Montana. Next, draft CFAs are evaluated in the context of landscape intactness and public–private ownership patterns. If CFAs have equivalent values after the sixth step, priority is

BOX 12.1. FOUR SPECIES CONCEPTS THAT BRIDGE SINGLE- AND
MULTIPLE-SPECIES MANAGEMENT

Flagship species are large, photogenic, historically significant, cuddly, cute, or of direct benefit to humans; they are the charismatic animals most likely to make people smile, feel goosebumps, and write a check for conservation (e.g., grizzly bear, manatee, wolf, polar bear). The concept does not even pretend to relate to a species' interactions or response to human perturbations. Rather, it is purely a strategic concept for raising public awareness and financial support for broad-based conservation action. In contrast, keystone species may not be particularly beautiful (flagship), or have wide space use (umbrella), or be especially sensitive to perturbations (indicators), but if they are lost from the system they could take with them many other species (Mills 2007). Selecting a focal or surrogate species for successful landscape conservation consists of identifying a flagship species, representing key social, political, and financial public values and feasibility, in combination with an umbrella, indicator, or keystone species, representing critical biological value. An example of a focal species that meets biological, social, political, and financial values is the westslope cutthroat trout, which can be contrasted with a tailed frog. Westslope cutthroat trout need large intact landscapes to complete their life cycle, but because this species is not federally endangered, private landowners are less threatened by projects that enhance its habitat. Anglers prize the fish because they are big and beautiful and have historic significance, resulting in strong financial and political support for conservation of the species and others that overlap its range. In contrast, although the tailed frog may be a more beneficial indicator species for cold and clean headwater streams, it does not qualify as a flagship species because of the challenges to building broad-based community, political, and financial support for a little-known species of frog.

given to landscapes with large tracts of native vegetation that are embedded within at least 50 percent public ownership or where private landowners own large parcels. Working with a private landowner who owns 5 hectares of land may take about the same amount of time as working with one who owns 5,000 hectares, yet the biological payoff is apt to be far greater with the latter. Ideally, ownership patterns within CFAs should be a public–private mix, with higher priority given to areas with greater public ownership (e.g., 50–90 percent).

The next filter involves assessing existing community-based conservation groups working within the identified CFA, such as watershed, place-based interest, or other nonprofit organizations, and is a key to conservation success. We elaborate in detail in the next section on the value of community-based partners to leverage investments and outcomes. Note that selecting CFAs solely on the premise of active community-based groups should not outweigh the former seven steps in prioritizing CFAs. The combination of high biological values with community values is the recipe for success.

At this point threats to CFAs, such as residential development and mining or other forms of resource development, become an important filter. Using threats as the first step in planning and prioritization typically leads to reactive rather than proactive conservation and poses the risk of bypassing conservation opportunities in intact landscapes. The CFA model addresses threats after previous filters have been used and all other biological, scale, and ownership factors are equal. At this point in the process, CFAs should be selected based on a high level of threat where the stressor has not yet occurred or is still at a biologically manageable level. The CFA model seeks to abate threats by delivering proactive conservation and working in more intact areas to address threats before they create irreversible resource damage.

The final step in the CFA approach is formal selection of CFAs. The Partners Program selected ten CFAs covering 23 percent of Montana's land ownership (12 percent public and 11 percent private). Three of these CFAs were developed using the grizzly bear as a focal species (box 12.2).

Selecting the right places to work using a biologically based, thorough, and systematic approach is critical to implementing community-based landscape conservation. We suggest that working in the right place using CFAs and other scientifically sound strategies constitutes perhaps 20 percent of a recipe for success. The next two sections, working with the right people and the right projects, constitute the remaining 80 percent of this new conservation paradigm and its value in practice.

The Social Framework: Working with the Right People

Conservation success depends heavily on the art of working with people. Here we explore the social underpinnings at the core of community-based conservation, address the need to shift from "biologist-centric" to "partner-centric" conservation planning and implementation, suggest the key qualities of a successful conservation practitioner and team, and illustrate our case with a project in one watershed of western Montana.

The task of defining effective community-based conservation is beset by numerous challenges, because success requires organic and innovative

BOX 12.2. AN EXAMPLE OF BIOLOGICAL PLANNING AND CONSERVATION DESIGN
IN THE WESTERN UNITED STATES

Grizzly bears in the Northern Continental Divide Ecosystem or Crown of the Continent Ecosystem are an excellent example of biological planning through selection of a focal species. The Crown of the Continent encompasses more than 4
million hectares (10 million acres) and includes Glacier National Park and the
Bob Marshall Wilderness. Land ownership within the ecosystem is primarily public, with private land buffering the east (Rocky Mountain Front), the south
(Blackfoot Watershed), and the west (Swan Valley). Ten communities are located within the landscape, with roughly 2,000 private landowners. The geography of the ecoregion may seem like the right scale to deliver grizzly bear conservation, but the U.S. Fish and Wildlife Service chose to divide it into three CFAs
with staff residing in each area for two reasons. First, on-the-ground conservation success typically result from establishing trust and credibility within communities, which takes time and is best accomplished when conservation practitioners become active community members. This process can take at least 10
years with hundreds of individual landowners. A smaller scale provides a better
ratio for one-on-one contact. Finally, 4 million hectares is not practical for conservation action because of the complex relationship among habitats, ownership patterns, and focal species requirements.

strategies for diverse situations and participants (Brosius et al. 1998; Agrawal and Gibson 1999; Kumar 2005). A defined theoretical framework for
community-based conservation becomes an ever-moving target based on
place, purpose, participants, goals, and activities. Still, efforts are being
made to understand and define this new style of natural resource management. Key ingredients of community-based conservation are inclusiveness
and diversity of participants (i.e., "coalitions of the unalike"); an emphasis
on collaborative and consensus-based process; innovative approaches to intractable conflicts, and local, regional, or place-based characteristics in
terms of scale and broad public–private partnerships with opportunities to
learn from one another (Cestero 1999; Wondolleck and Yaffee 2000; Snow
2001).

Experiences in the Blackfoot watershed of west-central Montana
provide an important lens with which to view key ingredients of
community-based conservation and to emphasize a crucial shift: away from
"biologist-centric" conservation to "partner-centric" conservation. We define biologist-centric conservation as prioritization of conservation actions

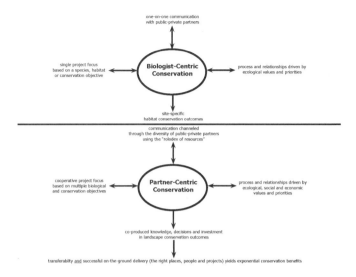

one-on-one communication
with public-private partners

single project focus
based on a species, habitat
or conservation objective

**Biologist-Centric
Conservation**

process and relationships driven by
ecological values and priorities

site-specific
habitat conservation outcomes

communication channeled
through the diversity of public-private partners
using the "rolodex of resources"

cooperative project focus
based on multiple biological
and conservation objectives

**Partner-Centric
Conservation**

process and relationships driven by
ecological, social and economic
values and priorities

co-produced knowledge, decisions and investment
in landscape conservation outcomes

transferabilty and successful on-the-ground delivery (the right places, people and projects) yields exponential conservation benefits

FIGURE 12.1. Schematic of the primary differences between biologist-centric and partner-centric approaches to conservation.

based on science, with the biologist and the resource of concern at the center of the decision-making process (fig. 12.1).

The emphasis is on quantifiable and verifiable empirical methods to define natural resource challenges and provide solutions. On the ground this means that goals are established based on agency mandates, with the biologist providing the lead role in communicating with private landowners and other partners. Typically, projects focus on a specific species-, habitat-, or agency-driven process for strategic conservation. In the Rocky Mountain West, for example, many public agencies and nonprofit organizations have at the core of their mission a mandate to protect threatened and endangered species. The result is a project goal that is driven by a species and its habitat needs, without the appropriate links across biophysical and social scales to leverage conservation opportunities and actions. Although using focal species is an appropriate nexus for prioritization, basing decisions on the knowledge base of the biologist can produce partial or incomplete outcomes for conservation, leading to missed opportunities to build relationships with private landowners for future conservation, restoration, and management activities on intermingled private and public lands.

Partner-centric conservation by nature is multidimensional in both process and outcome. From the outset, diverse private and public partners sit at the decision-making table, representing various interests, values, and skill sets and providing a range of technical and funding resources. Projects are cooperative in nature, focusing on multiple scales, biological and social, requiring a coproduced investment in the conservation outcome. Partners capitalize on the inherent link between the sustainability of natural resources being conserved and the communities that are tied to the resource. Although the process is organic, sometimes messy, two-steps-forward-one-step-back, the activities are replicated, with outputs leveraged and post-project outcomes exponentially multiplied. In the end, it becomes a reimagining of collaborative decision making that joins scientific expertise by biologists, academics, professionals, and consultants with community concerns, local traditions, and perspectives (Brick and Weber 2001).

The partner-centric model of conservation involves developing a team, or what Blackfoot rancher Jim Stone calls a "Rolodex of partners and resources," to implement conservation. In this model, private landowners and community members are recognized as critical participants in the process, and the role of the biologist is to help from behind instead of lead from the front. Successful projects are completed by finding landowner leaders or champions who are well respected, credible, and willing to initiate activities with their neighbors. Transferability of a conservation legacy between generations is greatly enhanced with this approach. Generational landowners typically have greater success in initiating momentum for conservation because they have the trust and credibility of long-standing friends and neighbors. However, as communities continue to cooperatively prioritize management, newer landowners are also pivotal to the success of many of these projects.

The landowner-led process is critical to producing long-term conservation outcomes. The value of this peer-to-peer relationship is often underestimated by conservation practitioners. Private landowners are the current and future land stewards; they are intimately tied to the resource, both its productivity and its limitations, and their place in the community allows them to communicate with neighbors about the importance of a project. The rural West is changing rapidly, its farms and ranches replaced by subdivisions, its small towns growing fast, and its very vastness and wildness challenged in many ways. Private landowners managing traditional working landscapes in the West are themselves decreasing in number with changing economics, demographics, and lifestyle preferences, resulting in growth and development by a mobile and wealthier population eager to become western (Riebsame 2001). The importance of generational landowners remaining active and productive on the landscape carries a profound weight and voice with policymakers in Washington, D.C., which is critical to acquiring funding and project support. When practitioners see their role in part as keeping private landowners with working landscapes in business, community-based conservation becomes not just a practice but an innovative partnership linking ecology, economy, and social viability.

The most successful conservation teams include a project coordinator with exceptional social and problem-solving skills (Low 2003). This person provides technical support and expertise to complete a project and has experience with on-the-ground delivery, providing support to the landowner leader and other community members. In many biologist-centric projects, new practitioners are expected to quickly meet organizational mandates without taking the time to understand the social landscape. These well-meaning practitioners fail to grasp the relationship between the community and its historical, ecological, social, and political heritage. Although there is no magic rule, we believe it takes at least 2 years to grasp the social landscape and 5 years to build the trust and credibility necessary to deliver community-based landscape conservation. During this critical period, it is important to make deposits into the community before making withdrawals. Deposits are often tied to community events and may include participation in recreational, civic, educational, and athletic activities. Practitioners should be available to private landowners outside normal working hours because community members volunteer their time attending meetings and implementing projects.

A successful project coordinator needs support from mid- to upper-level management within his or her agency. Agency decision makers benefit from hiring staff with diverse skills and expertise and empowering their staff with authority and flexibility. Project coordinators must be committed and caring people who are driven by results but are patient and persistent (Low 2003; table 12.2).

The successful project coordinator also must be a critical thinker with a bias for action, possess both book smarts and street smarts, and be adept as an institutional deal maker (Low 2003; table 12.2). Success requires teamwork through partners who are visionary, seeking solutions when seemingly intractable problems arise. Teams are enhanced by the inclusion of passionate, engaging communicators who can translate the project's value and details to stakeholders, decision makers, funders, and policymakers. Administrators must provide support related to maintaining contacts, grants and contracts, and budget oversight.

The Blackfoot Challenge in west-central Montana is a shining example of community-based landscape conservation (box 12.3).

TABLE 12.2. Key traits for hiring competent locally based conservation practitioners.

Key Trait	Description
Alignment with core values	Integrity beyond reproach; innovation and excellence; commitment to people and to the future.
Composure	Cool under pressure and handles stress; not knocked off balance by the unexpected; does not show frustration when resisted or blocked.
Deals well with ambiguity	Effectively copes with change; shifts gears; can decide and act without having the total picture; comfortably handles risk and uncertainty.
Drives for results	Bottom-line oriented; steadfastly pushes self and others for results; takes initiative to make concrete results happen; a deal maker.
Interpersonal savvy	Relates well to all kinds of people; builds constructive and effective relationships; displays diplomacy and tact.
Learns on the fly	Learns quickly when facing new problems; open to change; analyzes successes and failures for clues to improvement; seeks solutions.
Partnership oriented	Understands how to build a partnership for clearly defined results; active listener; collaborative; recognizes value of distinct strengths; openly shares credit.
Patience	Tolerant with people; sensitive to due process and proper pacing; tries to understand people and situations before judging and acting.
Perseverance	Pursues everything with energy, drive, and a need to finish; seldom gives up before finishing, especially when facing resistance or setbacks.
Political savvy	Maneuvers through complex political situations; anticipates landmines and plans approach accordingly; a "maze-bright" person.
Sizes up people	Good judge of talent; articulates people's strengths and limitations and anticipates what they will do in situations.
Strategic thinker	Crafts competitive and breakthrough strategies; can hold on to vision; sets aside the trivial and focuses on the critical.

BOX 12.3. THE BLACKFOOT CHALLENGE: AN EXAMPLE OF COMMUNITY-BASED
LANDSCAPE CONSERVATION IN NORTHWEST MONTANA

In the Blackfoot Valley located in west-central Montana, the landowner-based
group the Blackfoot Challenge and its partners are practicing community-based
partner-centric conservation. In 2000, a team of public and private partners
formed a Conservation Strategies Committee to share information, leverage
technical and funding resources, and determine which areas in the 607,000-
hectare (1.5 million-acre) watershed were in need of protection and cooperative
conservation through geographic information system mapping and integration
of data and plans by partnering agencies. This forum consisted of private landowners; a corporate timber company; federal agencies such as the U.S. Fish and
Wildlife Service, the U.S. Forest Service, the Bureau of Land Management, and
Natural Resources Conservation Service; state agencies such as Montana Fish,
Wildlife and Parks, and the Department of Natural Resources and Conservation;
nonprofit conservation partners such as The Nature Conservancy, Five Valleys
Land Trust, Montana Land Reliance, and the Rocky Mountain Elk Foundation; and
counties. They agreed to focus their strategic efforts on the midelevation Plum
Creek Timber Company (PCTC) lands that were becoming increasingly attractive
for real estate because of their amenity values. These transitional lands formed
important biological, agricultural, and public use access and use connections
between the higher-elevation public lands and the lower private valley bottoms.
At the time, half of the watershed's private land ownership was held in fee by
PCTC. A cooperative relationship with PCTC, built over the past decade, led the
Blackfoot Challenge and The Nature Conservancy to initiate a landscape-level
initiative in 2003 to purchase 36,104 hectares (89,215 acres) of PCTC lands and
resell the lands to public and private interests based on a community-driven
disposition plan. Known as the Blackfoot Community Project, the effort was a
result of trust between diverse private and public partners including a global
conservation organization, a local watershed-based group, a corporate timber
company, federal and state wildlife and resource agencies, three counties, and
five rural communities, exemplifying landowner-driven conservation action. In
contrast to purchase of the lands and a disposition strategy developed by the
agencies and organization groups, project partners sought community input and
support in the project before finalizing fee title transfer. Public meetings were
held in each of the affected communities to determine the community's values
related to the PCTC lands in their backyard and future management priorities,
seek recommendations as to whether specific parcels should be resold to public
or private interests, and ask public and private landowners with adjacent PCTC
parcels to indicate whether they would be interested in purchasing the lands.

BOX 12.3. CONTINUED

The Blackfoot Community Project embodies the heart of community-based collaborative conservation, with an integrated decision-making process of local and scientific expertise with landowner leaders, leveraging of multiple funding sources, coordination and staff assistance from the Blackfoot Challenge and The Nature Conservancy, and support by other partners at the table to acquire and hold perpetual conservation easements. The project was also replicated through the development of an innovative project called the Blackfoot Community Conservation Area, involving community-based ownership of former PCTC lands and cooperative ecosystem management across a 16,592-hectare (41,000-acre) landscape of public and private lands in the heart of the Blackfoot watershed.

The Blackfoot Challenge, along with similar grassroots organizations across the West, including the Malpai Borderlands Group and Quivera Coalition (New Mexico), Wallowa Resources (Oregon), Tallgrass Prairie Alliance (Kansas), Nebraska Sandhills, Northwest Connections, Swan Ecosystem Center, and Ranchers Stewardship Alliance (Montana), are actively practicing conservation by successfully mixing science with community values. Every public and private partner has a voice, and their interests are valued and incorporated into final decisions. Efforts to preserve species or habitats are enriched by community investments in the process and perpetuation of actions for the future. Community-based conservation is not the end but a means to making better decisions, building bridges between agencies, organizations, and private landowners, expanding support and capacity to meet future challenges (Wondolleck and Yaffee 2000).

Partner-centric conservation is driven by an emphasis on social processes and the formation of the right team of people. This approach blurs the lines between public and private interests, local knowledge and technical expertise, and biological and socioeconomic values. Community-based landscape conservation is practiced when partners working in the right places on the right projects follow what Montana rancher David Mannix calls the 80/20 rule, focusing on the 80 percent shared in common versus the 20 percent that divides partners. Once partners build trust and credibility by working on the 80 percent, they are able to tackle the 20 percent. In the end, success is born not from the efforts of one person but rather through a conservation community.

Conservation Delivery: Designing the Right Projects

Biological planning is critical to implementing landscape conservation, but lasting success depends on effective delivery through working with the right team of people to design the right projects. Conservation organizations excel at producing strategic habitat plans. Files in federal and state offices are filled with planning documents that have yielded little success in on-the-ground implementation. Worse yet, shortcomings in traditional agency planning processes have resulted in a train wreck of acrimony and distrust among stakeholders. The politics of expertise, lack of transparency and accountability, and inconsistent responsiveness to public concerns and issues has made planning by agencies a superficial and top-down exercise (Cortner and Moote 1999; Cortner et al. 2001). The problem is an inability to translate the plan into conservation delivery, and it is most evident in cumbersome procedural guidelines, complicated technical policies, and onerous eligibility requirements. This, combined with the failed recognition of the importance of personal relationships, practitioner social skills, and community support in conservation delivery, derails implementation. We believe these failures can be prevented by working in the right place with the right people and promoting the four key principles of effective conservation delivery: protection, management, restoration, and education.

Protection

Attitudes and approaches about land protection are constantly evolving. Land protection in North America essentially began as a movement to protect parks, wilderness areas, wildlife refuges, and other forms of preserves (Noss and Cooperrider 1994). We define land protection as conserving land through various public–private ownership strategies for the benefit of future generations. In the broadest sense, tools and techniques include fee title acquisition, conservation easements, transfer of development rights, leases, zoning, purchase and leaseback or sellback arrangements, management agreements, bargain sales, transfers in trust, statutory easements and scenic easements, community ownership or grassbanks, and other mechanisms that partition equity in land. In the past two decades, conservation practitioners have focused primarily on fee title acquisition and conservation easements as the primary tools for protecting land in perpetuity.

Fee title acquisition has been an important tool over time to ensure that natural resources are retained for the public good. Depending on the agency, criteria and goals for acquisition come in many forms, from areas critical for the maintenance of threatened and endangered species to lands providing timber and cultural and recreational uses. There have been both proponents for and opponents against fee title acquisition. Proponents typically support acquisition because they share acquisition values or goals. Opponents have been private property advocates or others concerned with land management practices and decreasing budgets, imperiled species issues, top-down regulation and enforcement, and tax revenue implications. Today, public agencies and private nonprofit organizations continue to use fee title acquisition as a land protection tool, with a focus on the private lands that are most threatened by subdivision or extractive industry development.

The emergence of the conservation easement as a land protection tool in the 1970s expanded the conservation toolbox and has helped to reverse landowner opposition to land protection in some areas. Conservation easements are either voluntarily sold or donated by private landowners to a qualified easement-holding organization to protect conservation values for perpetuity by limiting certain uses such as development. Conservation easements can be a preferred alternative to fee title acquisition because the land remains in private ownership, and traditional land uses such as forestry and agriculture are not only permitted but encouraged. Additionally, conservation easements can be tailored to meet a private landowner's economic and estate planning objectives. Despite these positive developments, implementation of a comprehensive conservation easement program is challenging. Conservation practitioners must be honest with communities and landowners about the implications of conservation easements and ensure that they understand the long-term commitment. When a conservation easement is purchased or donated, the legal document becomes a binding agreement joining the private landowner and the public agency or nonprofit land trust together for joint stewardship of the property.

Management

Conservation delivery does not begin and end with habitat protection; rather, management is equally important because ecosystems naturally evolved with some level of disturbance. Before European settlement, North America experienced wildfires, grazing by wild ungulates, insect infestations, droughts, and floods. Over the past 100 years, many of these natural influences have been controlled or lost. Total grazing exclusion, fire suppression, and inconsistent invasive species management adversely affect soil health, plant and animal diversity, and plant vigor. The most progressive conservation practitioners are applying management treatments that mimic natural processes. Management projects can include controlling invasive species, grazing management, and improvements in water conservation.

Management projects are often cost-effective and nonthreatening to land stewards; technical and financial assistance is typically available from an array of public and private partners. One of the best ways to build support for land protection is to first develop a short-term management agreement to address a stewardship issue. Practitioners can find common ground with landowners by providing technical assistance on grazing plans and cost-share for fencing, livestock water, and weed control. These early victories build trust and credibility, which often lead to other conservation opportunities. Management projects also provide an excellent learning environment for both landowners and conservation professionals. Landowners can learn about the habitat needs of focal species, and biologists can learn about the economic realities of ranching, energy production, and timber harvest. Working on management projects together often sets the stage for restoration.

Restoration

Habitat restoration is part of landscape conservation because almost every major habitat type in North America has been altered by human development. Restoration projects can be implemented quickly and economically and are known as early winners because they provide immediate and measurable benefits to focal species. As with management, completion of a successful restoration project typically sets the stage for larger and more complicated projects. Private landowners are motivated by projects that are result-oriented, due to their strong stewardship and work ethic, generational ties, and economic connection to the land. Timing, trust, patience, and dialogue about shared goals and objectives are needed when discussing restoration projects with landowners. Private landowners prefer to work with public agencies that have a successful record in working cooperatively with private landowners on habitat restoration, especially those that minimize paperwork, empower field biologists to become active community members, and implement efficient projects.

Education

Education is the mechanism for transferring the success and knowledge of community-based landscape conservation to current and future generations. Here we highlight strategies that provide complementary components to developing a comprehensive "3Ps" conservation program. The first is conducting field tours to share on-the-ground successes and challenges and to obtain support of potential funders, agency decision makers, and policymakers. There is simply no substitute for face-to-face interaction in the field, providing participants with the opportunity to ask questions and trade dialogue. Another key strategy is to work with educational institutions to incorporate natural history and conservation topics into curriculum, perhaps through research projects tied to flagship species. We advocate that universities develop more courses focused on community-based landscape conservation and the integration of a more holistic science (Cleveland et al. 2009). Finally, community education through various media and mechanisms—local newspaper articles, Web sites, field tours, project brochures, public meetings, workshops, and neighbor networks—all play a role in proactively addressing otherwise contentious resource issues. Respectful listening is perhaps the most important aspect of successful community-based landscape conservation. We urge practitioners to further their education through leadership training, partnership, and dialogue with landowners, and by learning from mentors who have successfully implemented community-based conservation.

Monitoring and Evaluation:
A Fine Line between Success and Failure

Measuring success provides partners with a biological or social basis for modifying strategies that ultimately delineate the fine line between success and failure. We suggest that project-based and overall conservation success should be measured via an intricate tapestry of biological and socioeconomic indicators. Evaluations of the biological response of focal species (chap. 3) are integral to assessing success in delivering conservation programs. Under this approach, professionals set explicit objectives for populations and monitor the responses of species. These indicators should complement other scientifically based monitoring and adaptive management. Biological assessments also raise credibility and transparency with decision makers by providing tangible accomplishments in the form of species currency.

Less intuitive but equally important is the need for rigorous evaluation of societal benefits of conservation. Societal benefits can be quantitative and qualitative indicators of processes and outcomes. Important dimensions of assessment include inclusiveness and diversity of participants; collaborative and consensus-based processes; innovation and creative approaches to intractable conflicts; local, regional, or place-based characteristics in terms of scale; and broad public–private partnerships with opportunities to learn from one another. Evaluation also requires assessment of the agency or nonprofit group's role in building strong relationships, such as providing opportunities for field biologists to become community members, and establishing trust and credibility. Other indicators of a project's contribution to society include partner involvement as measured by the scope, diversity, financial commitment, longevity, partner synergy, and satisfaction; community sustainability and links to natural resource viability, exploring the population, tax base, school enrollment, home and property foreclosures, community growth opportunities, and employment figures with respect to the resource economy; political support, assessed through consistent feedback and interest from local, state, and national elected officials in projects, new funding initiatives, and unsolicited requests for site visits and briefings; and transferability through assessing the level of interest in a project from other developing watershed and landscape groups.

The Upshot: Energy Development and Community-Based Landscape Conservation

From a conservation perspective, this book paints a sobering yet informed appraisal of the role the West will play in providing energy security for the United States and Canada. This book is a wake-up call to those who reject prioritizing landscapes for conservation and instead continue to work in highly degraded landscapes because they deny the inevitable impacts of energy development. The economic downturn in 2008–2011 slowed the pace of development, but our eventual return to prosperity, and an ever-widening gap between supply and demand, ensure that the extent and severity of impacts will soon increase.

Clearly, the answer to development in the West is not "no" but rather "where," and our biological understanding to date indicates that energy development and wildlife are largely incompatible (chaps. 3–7). Human nature dictates that the American people value first their own financial and national security, but they also value wildlife and habitat. Therefore, they will look to conservation practitioners for solutions in permitting development in some landscapes and maintaining healthy and connected wildlife populations in others. With one resounding voice, the conservation community needs to indicate which landscapes are most valuable to wildlife if they expect their interests to be heard. To date, we have no such game plan. Explicit in any solution are additional impacts on wildlife populations, but we can reduce the losses that would otherwise occur. Most importantly, it is the people and communities—not governments—who will ultimately envision and maintain lasting conservation in priority landscapes. We propose the principles of community-based landscape conservation contained herein as a viable means to that end.

Landscapes typically fall into one of three categories that characterize the potential choices and tradeoffs in selecting CFAs, each of which is ripe with opportunities and challenges from ecological and social perspectives: biologically impoverished landscapes with limited wildlife values, where impacts of energy development are small; biologically rich landscapes where substantial development activities already exist, necessitating mitigation measures; and biologically rich landscapes largely free of development, which demand 100 percent protection from an energy footprint.

The Prairie Pothole Region (PPR) of northern Montana, commonly called the hi-line, provides an example of how landscapes have been prioritized in a portion of this state. The central portion of the PPR is a severely fragmented landscape dominated by tillage agriculture that has high potential for wind and natural gas development. This area is classified as a low priority for conservation because energy reserves could be exploited here with minimal impact. This biologically impoverished landscape is exactly the place we should be directing energy development with minimal environmental regulations and red tape. The northeast and north-central portion of the PPR in Montana is of intermediate priority because conservation efforts were initiated too late to avoid degradation. Biological values are a mix of intact grasslands, shrublands, and wetland complexes offset by an intermingling of tillage agriculture and oil, natural gas, and wind developments. Additional energy development can occur here if appropriate onsite and offsite mitigation measures are implemented. The Rocky Mountain Front, which lies at the western edge of the PPR of northern Montana, is a largely intact landscape with unparalleled wildlife values. The Front Range represents the highest of conservation priorities that warrants complete protection from development but has been targeted by industry for its world-class natural gas and wind resources. Incentive-based protection from development was achieved on the Front Range in 2007 when a coalition of government and conservation organizations collaborated with energy companies to permanently acquire and retire oil, gas, and mineral leases.

Conclusions

In contrast to traditional conservation efforts that focus on an individual species, project, or habitat, community-based landscape conservation is multidimensional, working across spatial, temporal, ecological, and social scales. We provide a ten-step approach (table 12.1) to selecting CFAs that, when combined with partner-centric delivery of effective projects, is the next frontier in conservation. We present this approach as a model for identifying priority landscapes and for leveraging conservation investments and outcomes throughout the West. We urge conservation groups and public agencies to hire staff and practitioners who have the right social skills (table 12.2) and give them the flexibility to live and work in these communities to establish trust and credibility with partners. Success will depend in part on the ability of practitioners to forge partnerships with industry, which represent a potential new and unprecedented opportunity to fund conservation (chap. 9). The CFAs provide industry with an obvious place to invest in off-site mitigation and an opportunity for partners to collaborate with industry rather than demonize them when they contribute to conservation in priority landscapes—a situation that only results in heated debate, acrimony, and lawsuits, all to the continued detriment of the resource. Lastly, we urge all partners to learn from research so we can refine management adaptively through time (chap. 11) and continue to engage the people living and working in these landscapes so our efforts are maintained over multiple generations.

LITERATURE CITED

Aastrup, P. 2000. Responses of West Greenland caribou to the approach of humans on foot. *Polar Research* 19:83–90.

Abbitt, R. J. F., J. M. Scott, and D. S. Wilcove. 2000. The geography of vulnerability: Incorporating species geography and human development patterns into conservation planning. *Biological Conservation* 96:169–75.

Adams, L. G., F. J. Singer, and B. W. Dale. 1995. Caribou calf mortality in Denali National Park, Alaska. *Journal of Wildlife Management* 59:584–94.

Agrawal, A., and C. C. Gibson. 1999. Enchantment and disenchantment: The role of community in natural resource conservation. *World Development* 27:629–49.

Alberta Energy. 2010. *Oil sands facts and statistics*. Edmonton: Government of Alberta. www.energy.alberta.ca/OilSands/791.asp. Retrieved September 17, 2010.

Alberta Transportation. 2008. *Alberta highways traffic volume history 1962–2008*. www.transportation.alberta.ca/Content/docType181/Production/HTVH1962-2008.pdf. Retrieved February 15, 2009.

Alberta Woodland Caribou Recovery Team. 2005. *Alberta woodland caribou recovery plan, 2004/05–2013/14*. Alberta Sustainable Resource Development, Fish and Wildlife Division, Edmonton, Alberta, Canada.

Albon, S. D., T. N. Coulson, D. Brown, F. E. Guinness, J. M. Pemberton, and T. H. Clutton-Brock. 2000. Temporal changes in key factors and key age groups influencing the population dynamics of female red deer. *Journal of Animal Ecology* 69:1099–110.

Aldridge, C. L., and M. S. Boyce. 2007. Linking occurrence and fitness to persistence: Habitat-based approach for endangered greater sage-grouse. *Ecological Applications* 17:508–26.

Aldridge, C. L., M. S. Boyce, and R. K. Baydack. 2004. Adaptive management of prairie grouse: How do we get there? *Wildlife Society Bulletin* 32:92–103.

Allen, T. F. H., and T. B. Starr, eds. 1982. *Hierarchy: Perspectives for ecological complexity*. Chicago: University of Chicago Press.

American Gas Association. 2005. *Natural gas: Balancing supply, demand and the environment*. White paper delivered at the Natural Gas: Balancing Supply, Demand and the Environment Forum, May 24, 2005, Washington, DC.

American Wind Energy Association. 1995. *Avian Interactions with wind energy facilities: A summary*. Prepared by Colson and Associates for AWEA, Washington, DC.

——. 2006. *Wind energy fast facts*. www.awea.org/newsroom/pdf/FastFacts2006 .pdf. Retrieved September 3, 2008.

——. 2009. AWEA Web site. Home page: www.awea.org/; June 2009 capacity data available at www.awea.org/publications/reports/2Q09.pdf; Average household use from www.awea.org/newsroom/releases/Call_for_Immediate _Extension_of_Key_RE_Incentive_042108.html; current U.S. wind energy capacity by state (updated June 27, 2009): www.awea.org/projects/.

American Wind and Wildlife Institute. 2009. *The Wind–Wildlife Mapping Initiative*. www.awwi.org/initiatives/mapping.aspx. Retrieved December 13, 2009.

Amo, L., P. Lopez, and J. Martin. 2006. Nature-based tourism as a form of predation risk affects body condition and health state of *Podarcis muralis* lizards. *Biological Conservation* 131:402–9.

Amstrup, S. C. 1978. *Activities and habitat use of pronghorns on Montana–Wyoming coal lands*. Proceedings of the Eighth Biennial Pronghorn Antelope Workshop. Jasper, AB, Canada: U.S. Fish and Wildlife Service.

Anderson, R., N. Neumann, J. Tom, W. P. Erickson, M. D. Strickland, M. Bourassa, K. J. Bay, and K. J. Sernka. 2004. *Avian monitoring and risk assessment at the Tehachapi Pass Wind Resource Area, California*. NREL/SR-500-36416. Cheyenne, WY: Western EcoSystems Technology, Inc.

Anderson, R., J. Tom, N. Neumann, W. P. Erickson, M. D. Strickland, M. Bourassa, K. J. Bay, and K. J. Sernka. 2005. *Avian monitoring and risk assessment at the San Gorgonio Wind Resource Area (Phase I and II field work)*. NREL/SR-500-38054. Cheyenne, WY: Western EcoSystems Technology, Inc.

Anderson, S. H., K. Mann, and H. H. Shugart. 1977. Effect of transmission-line corridors on bird populations. *American Midland Naturalist* 97:216–21.

Andrews, A. 1990. Fragmentation of habitat by roads and utility corridors: A review. *Australian Zoologist* 26:130–41.

Anthony, R. G., J. A. Estes, M. A. Ricca, A. K. Miles, and E. D. Forsman. 2008. Bald eagles and sea otters in the Aleutian archipelago: Indirect effects of trophic cascades. *Ecology* 89:2725–35.

Armbruster, P., and R. Lande. 1993. A population viability analysis for African elephant (*Loxodonta africana*): How big should reserves be? *Conservation Biology* 7:602–10.

Armstrong, S. 1995. Rare plants protect cape's water supplies. *New Scientist* 145:8.

Arnett, E. B., W. K. Brown, W. P. Erickson, J. K. Fiedler, B. L. Hamilton, T. H. Henry, A. Jain, et al. 2008. Patterns of bat fatalities at wind energy facilities in North America. *Journal of Wildlife Management* 72:61–78.

Arnett, E. B., D. B. Inkley, D. H. Johnson, R. P. Larkin, S. Manes, A. M. Manville, J. R. Mason, et al. 2007. *Impacts of wind energy facilities on wildlife and wildlife habitat*. Technical Review 07-1. Bethesda, MD: The Wildlife Society.

Arnett, E. B., M. Schirmacher, M. M. P. Huso, and J. P. Hayes. 2009. *Effectiveness of changing wind turbine cut-in speed to reduce bat fatalities at wind facilities*. Annual report submitted to the Bats and Wind Energy Cooperative. Bat Conservation International.

Arponen, A., H. Kondelin, and A. Moilanen. 2007. Area-based refinement for selection of reserve sites with the benefit-function approach. *Conservation Biology* 21:527–33.

Ashenhurst, A. R., and S. J. Hannon. 2008. Effects of seismic lines on the abundance of breeding birds in the Kendall Island Bird Sanctuary, Northwest Territories, Canada. *Arctic* 61:190–98.

Askins, R. A., F. Chávez-Ramírez, B. C. Dale, C. A. Haas, J. R. Herkert, F. L. Knopf, and P. D. Vickery. 2007. Conservation of grassland birds in North America: Understanding ecological process in different regions. *Ornithological Monographs* 64.

Auld, G., S. Bernstein, and B. Cashore. 2008a. The new corporate social responsibility. *Annual Review of Environment and Resources* 33:413–35.

Auld, G., L. H. Gulbrandsen, and C. L. McDermott. 2008b. Certification schemes and the impacts on forests and forestry. *Annual Review of Environment and Resources* 33:187–211.

Avian Power Line Interaction Committee. 1994. *Mitigating bird collisions with power lines: The state of the art in 1994*. Washington, DC: Edison Electric Institute.

AXYS Environmental Consulting and Penner and Associates. 1998. *Environmental effects report: Wildlife*. Prepared for Diavik Diamonds Project, Yellowknife, Northwest Territories.

Aydemir, A. 2008. Hydrocarbon potential of the Tuzgolu (Salt Lake) Basin, Central Anatolia, Turkey: A comparison of geophysical investigation results with the geochemical data. *Journal of Petroleum Science and Engineering* 61:33–47.

Baerwald, E. F. 2008. *Variation in the activity and fatality of migratory bats at wind energy facilities in southern Alberta: Causes and consequences*. M.S. thesis, University of Calgary, Calgary, Alberta, Canada.

Baerwald, E. F., G. H. D'Amours, B. J. Klug, and R. M. R. Barclay. 2008. Barotrauma is a significant cause of bat fatalities at wind turbines. *Current Biology* 18:695–6.

Baerwald, E. F., J. Edworthy, M. Holder, and R. M. R. Barclay. 2009. A large-scale mitigation experiment to reduce bat fatalities at wind energy facilities. *Journal of Wildlife Management* 73:1077–81.

Bailey, R. G. 1995. *Description of the ecoregions of the United States*. Miscellaneous Publication No. 1391. Ogden, UT: U.S. Department of Agriculture Forest Service.

Baker, W. L. 2011. Pre-Euroamerican and recent fire in sagebrush ecosystems. Number 12 in S. T. Knick and J. W. Connelly, eds. Greater sage-grouse: Ecology and conservation of a landscape species and its habitats. *Studies in Avian Biology*. sagemap.wr.usgs.gov/monograph.aspx. Retrieved September 10, 2010.

Ball, I. R. 2000. *Mathematical applications for conservation ecology: The dynamics of tree hollows and the design of nature reserves*. University of Adelaide, South Australia.

Ball, J. R., E. M. Bayne, and C. M. Machtans. 2009. Energy sector edge effects on songbird nest fate and nest productivity in the boreal forest of western Canada: A preliminary analysis. *Proceedings of Tundra to Tropics: Connecting Birds, Habitats and People*. Proceedings of the 4th international Partners in Flight conference.

Ball, I. R., and H. P. Possingham. 2000. *Marxan V1.8.2: Manual for marine reserve design using spatially explicit annealing*. www.uq.edu.au/marxan/index.html?page=77655. Retrieved November 20, 2009.

Balmford, A., J. L. Moore, T. Brooks, N. Burgess, L. A. Hansen, P. Williams, and C. Rahbek. 2001. Conservation conflicts across Africa. *Science* 291:2616–19.

Bartelmus, P. 1997. Measuring sustainability: Data linkage and integration. Pages 110–118 in B. Moldan and S. Billharz, eds. *Sustainability indicators: A report on the project on indicators of sustainable development*. Cheltenham, UK: Countryside Commission.

Barten, N. L., R. T. Bowyer, and K. J. Jenkins. 2001. Habitat use by female caribou: Tradeoffs associated with parturition. *Journal of Wildlife Management* 65:77–92.

Bartis, J. T., T. LaTourrette, L. Dixon, D. J. Peterson, and G. Cecchine. 2005. *Oil shale development in the United States: Prospects and policy issues*. Santa Monica, CA: RAND Corporation.

Bartoldus, C. C. 1999. *A comprehensive review of wetland assessment procedures: A guide for wetland practitioners*. St. Michaels, MD: Environmental Concern, Inc.

Bascompte, J., C. J. Melian, and E. Sala. 2005. Interaction strength combinations and the overfishing of a marine food web. *Proceedings of the National Academy of Sciences of the United States of America* 102:5443–7.

Bashkin, M., T. J. Stohlgren, Y. Otsuki, M. Lee, P. Evangelista, and J. Belnap. 2003. Soil characteristics and plant exotic species invasions in the Grand Staircase–Escalante National Monument, Utah, USA. *Applied Soil Ecology* 22:67–77.

Baskin, C. C., and J. M. Baskin, eds. 1998. *Seeds: Ecology, biogeography, and evolution of dormancy and germination*. New York: Academic Press.

Battin, J. 2004. When good animals love bad habitats: Ecological traps and the conservation of animal populations. *Conservation Biology* 18:1482–91.

Baxter, W., W. A. Ross, and H. Spaling. 2001. Improving the practice of cumulative effects assessment in Canada. *Impact Assessment and Project Appraisal* 19.

Baydack, R. K., and D. A. Hein. 1987. Tolerance of sharp-tailed grouse to lek disturbance. *Wildlife Society Bulletin* 15:535–9.

Bayne, E. M., S. Boutin, B. Tracz, and K. Charest. 2005a. Functional and numerical responses of ovenbirds (*Seiurus aurocapilla*) to changing seismic exploration practices in Alberta's boreal forest. *Ecoscience* 12:216–22.

Bayne, E. M., L. Habib, and S. Boutin. 2008. Impacts of chronic anthropogenic noise from energy-sector activity on abundance of songbirds in the boreal forest. *Conservation Biology* 22:1186–93.

Bayne, E. M., and K. A. Hobson. 2002. Apparent survival of male ovenbirds in fragmented and forested boreal landscapes. *Ecology* 83:1307–16.

Bayne, E. M., S. L. Van Wilgenburg, S. Boutin, and K. A. Hobson. 2005b. Modeling and field-testing of ovenbird (*Seiurus aurocapillus*) responses to boreal forest dissection by energy sector development at multiple spatial scales. *Landscape Ecology* 20:203–16.

Beck, J. L., K. P. Reese, J. W. Connelly, and M. B. Lucia. 2006. Movements and survival of juvenile greater sage-grouse in southeastern Idaho. *Wildlife Society Bulletin* 34:1070–78.

Beckers, J., Y. Alila, and A. Mtiraoui. 2002. On the validity of the British Columbia Forest Practices Code guidelines for stream culvert discharge design. *Canadian Journal of Forestry Research* 32:648–92.

Beierle, T. C., and R. J. Long. 1999. Chilling collaboration: The Federal Advisory Committee Act and stakeholder involvement in environmental decisionmaking. *Environmental Law Reporter* 39:10406–7.

Belnap, J., M. Reheis, and R. Reynolds. 2000. *Conditions favoring and retarding cheatgrass invasion of arid lands in the southwestern U.S.* esp.cr.usgs.gov/info/sw/cheatgrass/. Retrieved June 1, 2009.

Bennett, V. J., M. Beard, P. A. Zollner, E. Fernandez-Juricic, L. Westphal, and C. L. LeBlanc. 2009. Understanding wildlife responses to human disturbance through simulation modelling: A management tool. *Ecological Complexity* 6:113–34.

Bennington, J. P., R. S. Dressler and J. M. Bridges. 1982. The effect of hydrocarbon development on elk and other wildlife in lower Michigan. Pages 363–84 in P. J. Ravel, ed. *Land and water issues related to energy development*. Ann Arbor, MI: Ann Arbor Science.

Benson, M. H. 2009. Integrating adaptive management and oil and gas development: Existing obstacles and opportunities for reform. *Environmental Law Reporter* 39:10962–78.

Berger, J. 2004. The last mile: How to sustain long-distance migration in mammals. *Conservation Biology* 18:320–31.

———. 2007. Fear, human shields and the redistribution of prey and predators in protected areas. *Biology Letters* 3:620–3.

Berger, J., K. M. Berger, and J. Beckman. 2006a. *Wildlife and energy development: Pronghorn of the Upper Green River Basin: Year 1 summary*. Bronx, NY: Wildlife Conservation Society.

Berger, J., S. L. Cain, and K. M. Berger. 2006b. Connecting the dots: An invariant migration corridor links the Holocene to the present. *Biology Letters* 2:528–31.

Berger, K. M., J. Beckman, and J. Berger. 2007. *Wildlife and energy development:*

Pronghorn of the Upper Green River Basin—year 2 summary. Bronx, NY: Wildlife Conservation Society.

Berger, K. M., E. M. Gese, and J. Berger. 2008. Indirect effects and traditional trophic cascades: A test involving wolves, coyotes, and pronghorn. *Ecology* 89:818–28.

Bergerud, A. T., and J. P. Elliot. 1986. Dynamics of caribou and wolves in northern British Columbia. *Canadian Journal of Zoology* 64:1515–29.

Bergerud, A. T., and S. N. Luttich. 2003. Predation risk and optimal foraging trade-off in the demography and spacing of the George River Herd, 1958 to 1993. *Rangifer* 14:169–91.

Bergman, E. J., R. A. Garrott, S. Creel, J. J. Borkowski, R. Jaffe, and E. G. R. Watson. 2006. Assessment of prey vulnerability through analysis of wolf movements and kill sites. *Ecological Applications* 16:273–84.

Bergquist, E., P. Evangelista, T. J. Stohlgren, and N. Alley. 2007. Invasive species and coal bed methane development in the Powder River Basin, Wyoming. *Environmental Monitoring and Assessment* 128:381–94.

Best, L. B. 1986. Conservation tillage: Ecological traps for nesting birds? *Wildlife Society Bulletin* 14:308–17.

Bevanger, K., and P. G. Thingstad. 1988. The relationship of birds–constructions for transmission of electric energy. A survey of present knowledge. *Økoforsk Utredning* 1:1–133.

Bisson, I. A., L. K. Butler, T. J. Hayden, L. M. Romero, and M. C. Wikelski. 2009. No energetic cost of anthropogenic disturbance in a songbird. *Proceedings of the Royal Society B: Biological Sciences* 276:961–9.

Bjorge, R. R. 1987. Bird kill at an oil industry flare stack in northwest Alberta. *Canadian Field-Naturalist* 101:346–50.

Blackburn, W. H., R. W. Knight, and J. L. Schuster. 1982. Saltcedar influence on sedimentation in the Brazos River. *Journal of Soil and Water Conservation* 37:298–301.

Blancher, P., and J. Wells. 2005. *The boreal forest region: North America's bird nursery*. Boreal Songbird Initiative. www.borealbirds.org/resources/report-bsi-birdnursery.pdf. Retrieved March 7, 2008.

Blondel, J. 2008. On humans and wildlife in Mediterranean islands. *Journal of Biogeography* 35:509–18.

Boal, C. W., and R. W. Mannan. 1999. Comparative breeding ecology of Cooper's hawks in urban and exurban areas of southeastern Arizona. *Journal of Wildlife Management* 63:77–84.

Bollinger, E. K., P. B. Bollinger, and T. A. Gavin. 1990. Effects of hay-cropping on eastern populations of the bobolink. *Wildlife Society Bulletin* 18:142–50.

Borkowski, J. 2001. Flight behaviour and observability in human-disturbed sika deer. *Acta Theriologica* 46:195–206.

Borrvall, C., and B. Ebenman. 2006. Early onset of secondary extinctions in ecological communities following the loss of top predators. *Ecology Letters* 9:435–42.

Bottrill, M. C., L. N. Joseph, J. Carwardine, M. Bode, C. Cook, E. T. Game, H. Grantham, et al. 2008. Is conservation triage just smart decision making? *Trends in Ecology and Evolution* 23:649–54.

———. 2009. Finite conservation funds mean triage is unavoidable. *Trends in Ecology and Evolution* 24:183–4.

Boyce, M. S. 1992. Population viability analysis. *Annual Review of Ecology and Systematics* 23:481–506.

Boyce, M. S., and L. L. McDonald. 1999. Relating populations to habitats using resource selection functions. *Trends in Ecology and Evolution* 14:268–72.

Bradford, D. F., S. E. Franson, A. C. Neale, D. T. Heggem, G. R. Miller, and G. E. Canterbury. 1999. Bird species assemblages as indicators of biological integrity in Great Basin rangeland. *Environmental Monitoring and Assessment* 49:1–22.

Bradshaw, C. J. A., S. Boutin, and D. M. Hebert. 1997. Effects of petroleum exploration on woodland caribou in northeastern Alberta. *Journal of Wildlife Management* 61:1127–33.

———. 1998. Energetic implications of disturbance caused by petroleum exploration to woodland caribou. *Canadian Journal of Zoology* 76:1319–24.

Braun, C. E. 1998. Sage grouse declines in western North America: What are the problems? *Proceedings of the Western Association of State Fish and Wildlife Agencies* 78:139–56.

Breiman, L. 2001. Random forests. *Machine Learning* 45:5–32.

Brennan, L. A., and W. P. Kuvlesky. 2005. North American grassland birds: An unfolding conservation crisis? *Journal of Wildlife Management* 69:1–13.

Brick, P., and E. P. Weber, eds. 2001. *Will rain follow the plow? Unearthing a new environmental movement.* Washington, DC: Island Press.

British Columbia Oil and Gas Commission. 2004. *Geophysical guidelines for the Muskwa-Kechika Management Area.* Fort St. John, BC, Canada.

Brodeur, V., J. P. Ouellet, R. Courtois, and D. Fortin. 2008. Habitat selection by black bears in an intensively logged boreal forest. *Canadian Journal of Zoology* 86:1307–16.

Bromley, M. 1985. *Wildlife management implications of petroleum exploration and development in wildland environments.* Washington, DC: U.S. Forest Service.

Brooke, C. 2008. Conservation and adaptation to climate change. *Conservation Biology* 22:1471–6.

Brooks, T. M., R. A. Mittermeier, C. G. Mittermeier, G. A. B. da Fonseca, A. B. Rylands, W. R. Konstant, P. Flick, et al. 2002. Habitat loss and extinction in the hotspots of biodiversity. *Conservation Biology* 16:909–23.

Brosius, J. P., A. L. Tsing, and C. Zerner. 1998. Representing communities: Histories and politics of community-based natural resource management. *Society and Natural Resources* 2:157–68.

Brothers, T. S., and A. Spingarn. 1992. Forest fragmentation and alien plant invasion of central Indiana old-growth forests. *Conservation Biology* 6:91–100.

Brown, C. D., and C. Boutin. 2009. Linking past land use, recent disturbance, and dispersal mechanism to forest composition. *Biological Conservation* 142:1647–56.

Brown, J. W. 2006. *Eco-logical: An ecosystem approach to developing infrastructure projects.* Washington, DC: Office of Project Development and Environmental Review, Federal Highway Administration.

Brown, W. K., and B. L. Hamilton. 2002. *Bird and bat interactions with wind turbines Castle River Wind Farm, Alberta.* Calgary, AB: VisionQuest Windelectric, Inc.

———. 2006. *Bird and bat monitoring at the McBride Lake Wind Farm, Alberta, 2003–2004.* Calgary, AB: Vision Quest Windelectric, Inc.

Bull, K. A., K. W. Stolte, and T. J. Stohlgren. 1998. *Forest health monitoring: Vegetation pilot field methods guide.* Washington, DC: U.S. Department of Agriculture, Forest Service.

Burba, E. A., G. D. Schnell, and J. A. Grzybowski. 2008. *Post-construction avian/bat fatality study for the Blue Canyon II Wind Power Project, Oklahoma: Summary of preliminary findings for 2006–2007.* Interim report to Horizon Wind Energy, Sam Noble Oklahoma. Museum of Natural History, University of Oklahoma, Norman.

Bureau of Land Management (BLM). 2003a. *Final environmental impact statement and proposed plan amendment for the Powder River Basin oil and gas project.* Cheyenne, WY: BLM State Office.

———. 2003b. *Montana statewide final oil and gas EIS amendment of the Powder River and Billings Resource Management Plans.* Billings, MT: BLM State Office.

———. 2005. *Final programmatic environmental impact statement on wind energy development on BLM administered land in the western United States.* Washington, DC: U.S. Department of the Interior, BLM.

———. 2009. *Oil shale and tar sands programmatic environmental impact statement.* Washington, DC: U.S. Department of the Interior.

Burgin, S. 2008. BioBanking: An environmental scientist's view of the role of biodiversity banking offsets in conservation. *Biodiversity and Conservation* 17:807–16.

Burris, R. K., and L. W. Canter. 1997. Cumulative impacts are not properly addressed in environmental assessments. *Environmental Impact Assessment Review* 17:5–18.

Busch, D. E., and S. D. Smith. 1995. Mechanisms associated with decline of woody species in riparian ecosystems of the southwestern U.S. *Ecological Monographs* 65:347–70.

Caldwell, L. K., ed. 2002. *The National Environmental Policy Act: Judicial misconstruction, legislative indifference, and executive neglect.* College Station: Texas A & M University Press.

California Energy Commission. 2009. *On-line annotated bibliography of avian interactions with utility structures.* www.energy.ca.gov/research/environmental/avian_bibliography/index.html. Retrieved September 22, 2009.

Cameron, R. D., D. J. Reed, J. R. Dau, and W. T. Smith. 1992. Redistribution of calving caribou in response to oil-field development on the Arctic slope of Alaska. *Arctic* 45:338–42.

Cameron, R. D., W. T. Smith, R. G. White, and B. Griffith. 2005. Central Arctic caribou and petroleum development: Distributional, nutritional, and reproductive implications. *Arctic* 58:1–9.

Canadian Wind Energy Association. 2009. CWEA Web site. Home page: www .canwea.ca/; current Canadian wind energy capacity by province: www.canwea .ca/farms/index_e.php. Retrieved July 15, 2009.

Canter, L., ed. 1996. *Environmental impact assessment*, 2nd ed. New York: McGraw-Hill.

Carman, J. G., and J. D. Brotherson. 1982. Comparisons of sites infested and not infested with saltcedar (*Tamarix pendandra*) and Russian olive (*Elaeagnus angustifolia*). *Weed Science* 30:360–4.

Carpenter, J., C. Aldridge, and M. S. Boyce. 2010. Sage-grouse habitat selection during winter in Alberta. *Journal of Wildlife Management*.

Carson, R., ed. 1962. *Silent spring*. New York: Houghton Mifflin.

Carter, T. C., M. A. Menzel, and D. A. Saugey. 2003. Population trends of solitary foliage-roosting bats. Pages 41–47 in T. J. O'Shea and M. A. Bogan, eds. *Monitoring trends in bat populations of the United States and territories: Problems and prospects*. Biological Resources Discipline, Information, and Technology Report USGS/BRD/ITR-2003-003. Washington, DC: U.S. Geological Survey.

Caughley, G., ed. 1977. *Analysis of vertebrate populations*. New York: Wiley and Sons.

Ceballos, G., and P. R. Ehrlich. 2002. Mammal population losses and the extinction crisis. *Science* 296:904–7.

Cerutti, P. O., R. Nasi, and L. Tacconi. 2008. Sustainable forest management in Cameroon needs more than approved forest management plans. *Ecology and Society* 13:36.

Cestero, B. 1999. *Beyond the hundredth meeting: A field guide to collaborative conservation on the West's public lands*. Tucson, AZ: Sonoran Institute.

Chambers, R., ed. 1997. *Whose reality counts? Putting the first last*. London: Intermediate Technology Productions.

Chan-McLeod, A. C. A., R. G. White, and D. F. Holleman. 1994. Effects of protein and energy intake, body condition, and season on nutrient partitioning and milk production in caribou and reindeer. *Canadian Journal of Zoology* 72:938–47.

Chasko, G. G., and J. E. Gates. 1982. Avian habitat suitability along a transmission line corridor in an oak-hickory forest region. *Wildlife Monographs* 82.

Chen, Z. H., K. Osadetz, H. Y. Gao, P. Hannigan, and C. Watson. 2000. Characterizing the spatial distribution of an undiscovered hydrocarbon resource: The Keg River Reef play, Western Canada Sedimentary Basin. *Bulletin of Canadian Petroleum Geology* 48:150–62.

Chornesky, E. A., and J. M. Randall. 2003. The threat of invasive alien species to biological diversity: Setting a future course. *Annals of the Missouri Botanical Garden* 90:67–76.

Christensen, E. M. 1962. The rate of naturalization of *Tamarix* in Utah. *American Midland Naturalist* 6851–57.

Christian, J. M., and S. D. Wilson. 1999. Long-term ecosystem impacts of an introduced grass in the northern Great Plains. *Ecology* 80:2397–407.

Clark, W. C. 2007. Sustainability science: A room of its own. *Proceedings of the National Academy of Sciences of the United States of America* 104:1737–8.

Clemmons, J. R., and R. Buchholz, eds. 1997. *Linking conservation and behaviour*. Cambridge: Cambridge University Press.

Cleveland, S. M., J. Merkle, R. Stutzman, G. Neudecker, J. Stone, and D. Naugle. 2009. Lessons from the field: A hands-on crash course in landscape conservation. *The Wildlife Professional* 3:40–41.

Clevenger, A. P., B. Chruszczc, and K. E. Gunson. 2003. Spatial patterns and factors influencing small vertebrate fauna road-kill aggregations. *Biological Conservation* 109:15–26.

Cleverly, J. R., S. D. Smith, A. Sala, and D. A. Devitt. 1997. Invasive capacity of *Tamarix ramosissima* in a Mojave Desert floodplain: The role of drought. *Oecologia* 111:12–8.

Cohen, J. 1994. The earth is round ($P < .05$). *American Psychologist* 49:997–1003.

Committee on the Status of Endangered Wildlife in Canada. 2002. *COSEWIC assessment and update status report on the woodland caribou,* Rangifer tarandus *caribou, in Canada*. Ottawa: COSEWIC, Environment Canada.

Conacher, A. J. 1994. The integration of land-use planning and management with environmental impact assessment: Some Australian and Canadian perspectives. *Impact Assessment* 4:347–73.

Confer, J. L., and S. M. Pascoe. 2003. Avian communities on utility rights-of-ways and other managed shrublands in the northeastern United States. *Forest Ecology and Management* 185:193–205.

Connelly, J. W., and C. E. Braun. 1997. Long-term changes in sage grouse *Centrocercus urophasianus* populations in western North America. *Wildlife Biology* 3:229–34.

Connelly, J. W., A. D. Apa, R. B. Smith, and K. P. Reese. 2000a. Effects of predation and hunting on adult sage grouse *Centrocercus urophasianus* in Idaho. *Wildlife Biology* 6:227–32.

Connelly, J. W., M. A. Schroeder, A. R. Sands, and C. E. Braun. 2000b. Guidelines to manage sage grouse populations and their habitats. *Wildlife Society Bulletin* 28:967–85.

Connelly, J. W., S. T. Knick, M. A. Schroeder, and S. J. Stiver. 2004. *Conservation assessment of greater sage-grouse and sagebrush habitats*. Unpublished report, Western Association of Fish and Wildlife Agencies.

Connelly, J. W., E. T. Rinkes, and C. E. Braun. 2011. Characteristics of greater sage-grouse habitats: A landscape species at micro and macro scales. Number 4 in

S. T. Knick and J. W. Connelly, eds. Greater sage-grouse: Ecology and conservation of a landscape species and its habitats. *Studies in Avian Biology*. sagemap .wr.usgs.gov/monograph.aspx. Retrieved January 5, 2010.

Cook, J. G., B. K. Johnson, R. C. Cook, R. A. Riggs, T. Delcurto, L. D. Bryant, and L. L. Irwin. 2004. Effects of summer–autumn nutrition and parturition date on reproduction and survival of elk. *Wildlife Monographs* 155.

Copeland, H. E., K. E. Doherty, D. E. Naugle, A. Pocewicz, and J. M. Kiesecker. 2009. Mapping oil and gas development potential in the US intermountain West and estimating impacts to species. *PLoS One* 4:e7400.

Copeland, H., J. Ward, and J. M. Kiesecker. 2007. Threat, cost, and biological value: Prioritizing conservation within Wyoming ecoregions. *Journal of Conservation Planning* 3:1–16.

Cortner, H. J., and M. A. Moote, eds. 1999. *The politics of ecosystem management*. Washington, DC: Island Press.

Cortner, H. J., S. Burns, L. R. Clark, W. H. Sanders, G. Townes, and M. Twarkins. 2001. Governance and institutions: Opportunities and challenges. *Journal of Sustainable Forestry* 12:65–96.

Coulson, T., J. M. Gaillard, and M. Festa-Bianchet. 2005. Decomposing the variation in population growth into contributions from multiple demographic rates. *Journal of Animal Ecology* 74:789–801.

Council on Environmental Quality. 2000. Protection of the environment under the National Environment Policy Act. *Code of Federal Regulations*, title 40, sec. 5 (1500–17).

Courtois, R., and J. P. Ouellet. 2007. Modeling the impact of moose and wolf management on persistence of woodland caribou. *Alces* 43:13–27.

Courtois, R., J. P. Ouellet, A. Gingras, C. Dussault, L. Breton, and J. Maltais. 2003. Historical changes and current distribution of caribou, *Rangifer tarandus*, in Quebec. *Canadian Field-Naturalist* 117:399–414.

Crawford, J. A., R. A. Olson, N. E. West, J. C. Mosley, M. A. Schroeder, T. D. Whitson, R. F. Miller, M. A. Gregg, and C. S. Boyd. 2004. Ecology and management of sage-grouse and sage-grouse habitat. *Journal of Range Management* 57:2–19.

Cronin, M. A., W. B. Ballard, J. D. Bryan, B. J. Pierson, and J. D. McKendrick. 1998. Northern Alaska oil fields and caribou: A commentary. *Biological Conservation* 83:195–208.

Cronin, M. A., H. A. Whitlaw, and W. B. Ballard. 2000. Northern Alaska oil fields and caribou. *Wildlife Society Bulletin* 28:919–22.

Crooks, J. A. 2005. Lag times and exotic species: The ecology and management of biological invasions in slow-motion. *Ecoscience* 12:316–29.

Cubbage, F. W., and D. H. Newman. 2006. Forest policy reformed: A United States perspective. *Forest Policy and Economics* 9:261–73.

Cutler, D. R., T. C. Edwards, K. H. Beard, A. Cutler, and K. T. Hess. 2007. Random forests for classification in ecology. *Ecology* 88:2783–92.

Dahle, B., E. Reimers, and J. E. Coleman. 2008. Reindeer (*Rangifer tarandus*) avoidance of a highway as revealed by lichen measurements. *European Journal of Wildlife Research* 54:27–35.

Dale, B. C., P. A. Martin, and P. S. Taylor. 1997. Effects of hay management on grassland songbirds in Saskatchewan. *Wildlife Society Bulletin* 25:616–26.

Dale, B. C., T. S. Wiens, and E. L. E. Hamilton. 2009. Abundance of three grassland songbirds in an area of natural gas infill drilling in Alberta, Canada. Pages 194–204 in *Proceedings of the 4th International Partners in Flight Conference—Tundra to Tropics: Connecting Birds, Habitats and People*.

Dale, V. H., S. Brown, R. A. Haeuber, N. T. Hobbs, N. Huntly, R. J. Naiman, W. E. Riebsame, M. G. Turner, and T. J. Valone. 2000. Ecological principles and guidelines for managing use of the land. *Ecological Applications* 10:639–70.

D'Antonio, C. M., ed. 2000. *Fire, plant invasions, and global changes*. Washington, DC: Island Press.

D'Antonio, C. M., T. L. Dudley, and M. C. Mack, eds. 1999. *Disturbance and biological invasions: Direct effects and feedback*. Amsterdam: Elsevier.

D'Antonio, C. M., and S. E. Hobbie, eds. 2005. *Plant species effects on ecosystem processes: Insights from invasive species*. Sunderland, MA: Sinauer.

D'Antonio, C. M., J. M. Levine, and M. Thomsen. 2001. Ecosystem resistance to invasion and the role of propagule supply: A California perspective. *Journal of Mediterranean Ecology* 2:233–45.

D'Antonio, C. M., and L. A. Meyerson. 2002. Exotic plant species as problems and solutions in ecological restoration: A synthesis. *Restoration Ecology* 10:703–13.

D'Antonio, C. M., and P. M. Vitousek. 1992. Biological invasions by exotic grasses, the grass/fire cycle, and global change. *Annual Review of Ecological Systems* 23:63–87.

Davey, L. H., J. L. Barnes, C. L. Horvath, and A. Griffiths. 2001. *Addressing cumulative environmental effects: Sectoral and regional environmental assessment*. Proceedings of Cumulative Environmental Effects Management, Tools and Approaches. Alberta Society of Professional Biologists.

Davies, R. G., C. D. L. Orme, V. Olson, G. H. Thomas, S. G. Ross, T. S. Ding, P. C. Rasmussen, et al. 2006. Human impacts and the global distribution of extinction risk. *Proceedings of the Royal Society B: Biological Sciences* 273:2127–33.

Davis, S. K. 2004. Area sensitivity in grassland passerines: Effects of patch size, patch shape, and vegetation structure on bird abundance and occurrence in southern Saskatchewan. *Auk* 121:1130–45.

Davis, S. K., and D. C. Duncan. 1999. Grassland songbird occurrence in native and crested wheatgrass pastures of southern Saskatchewan. *Studies in Avian Biology* 19:211–8.

De'ath, G., and K. E. Fabricius. 2000. Classification and regression trees: A powerful yet simple technique for ecological data analysis. *Ecology* 81:3178–92.

DeCesare, N. J., M. Hebblewhite, H. S. Robinson, and M. Musiani. 2009. Endangered, apparently: The role of apparent competition in endangered species conservation. *Animal Conservation* online.

DeLuca, T. H., G. H. Aplet, and B. Wilmer. 2008. *The unknown trajectory of forest restoration: A call for ecosystem monitoring.* Washington, DC: The Wilderness Society.

de Lucas, M., G. F. E. Janss, and M. Ferrer. 2005. A bird and small mammal BACI and IG design studies in a wind farm in Malpica (Spain). *Biodiversity and Conservation* 14:3289–303.

Derby, C., A. Dahl, W. Erickson, K. Bay, and J. Hoban. 2007. *Post-construction monitoring report for avian and bat mortality at the NPPD Ainsworth Wind Farm.* Unpublished report prepared for Nebraska Public Power District by Western EcoSystems Technology, Inc., Cheyenne, WY.

Desholm, M. 2009. Avian sensitivity to mortality: Prioritising migratory bird species for assessment at proposed wind farms. *Journal of Environmental Management* 90:2672–9.

Desmet, P., and R. Cowling. 2004. Using the species–area relationship to set baseline targets for conservation. *Ecology and Society* 9:11–20.

Devereux, C. L., M. J. H. Denny, and M. J. Whittingham. 2008. Minimal effects of wind turbines on the distribution of wintering farmland birds. *Journal of Applied Ecology* 45:1689–94.

Diavik Diamond Mines Incorporated. 2008. *2007 wildlife monitoring program.* Yellowknife, Northwest Territories, Canada: Diavik Diamond Mines, Inc.

Diaz, M., J. C. Illera, and D. Hedo. 2001. Strategic environmental assessment of plans and programs: A methodology for estimating effects on biodiversity. *Environmental Management* 28:267–79.

Dinerstein, E., G. Powell, D. Olson, E. Wikramanayake, R. Abell, C. Loucks, E. Underwood, et al. 2000. A workbook for conducting biological assessments and developing biodiversity visions for ecoregion-based conservation. World Wildlife Fund, Conservation Science Program.

DiTomaso, J. M. 1998. Impact, biology, and ecology of saltcedar (*Tamarix* spp.) in the southwestern United States. *Weed Technology* 12:326–36.

DiTomaso, J. M., G. B. Kyser, and C. B. Pirosko. 2003. Effect of light and density on yellow starthistle (*Centaurea solstitialis*) root growth and soil moisture use. *Weed Science* 51:334–41.

Dixon, J., and B. E. Montz. 1995. From concept to practice: Implementing cumulative impact assessment in New Zealand. *Environmental Management* 19:445–56.

Doherty, K. E. 2008. *Sage-grouse and energy development: Integrating science with conservation planning to reduce impacts.* Ph.D. dissertation. University of Montana, Missoula.

Doherty, K., D. E. Naugle, H. Copeland, A. Pocewicz, and J. M. Kiesecker. 2011. Energy development and conservation tradeoffs: Systematic planning for sage-grouse in their eastern range. Number 22 in S. T. Knick and J. W. Connelly, eds. Greater sage-grouse: Ecology and conservation of a landscape species and its habitats. *Studies in Avian Biology.* sagemap.wr.usgs.gov/monograph .aspx. Retrieved September 17, 2010.

Doherty, K. E., D. E. Naugle, and J. S. Evans. 2010a. A currency for offsetting energy development impacts: Horse-trading sage-grouse on the open market. *PLoS One* 5:e10339.

Doherty, K. E., D. E. Naugle, B. L. Walker, and J. M. Graham. 2008. Greater sage-grouse winter habitat selection and energy development. *Journal of Wildlife Management* 72:187–95.

Doherty, K. E., J. D. Tack, J. S. Evans, and D. E. Naugle. 2010b. *Mapping breeding densities of greater sage-grouse: A tool for range-wide conservation planning*. Unpublished report, Bureau of Land Management, Washington, DC. Geo databases and report accessible at conserveonline.org/workspaces/sagegrouse. Retrieved September 17, 2010.

Doncaster, C. P., T. Micol, and S. P. Jensen. 1996. Determining minimum habitat requirements in theory and practice. *Oikos* 75:335–9.

Doremus, H. 2001. Adaptive management, the Endangered Species Act, and the institutional challenges of "New Age" environmental protection. *Washburn Law Journal* 41:50–89.

Dreher, R. G. 2005. *NEPA under siege: The political assault on the National Environmental Policy Act*. Washington, DC: Georgetown Environmental Law and Policy Institute, Georgetown University.

Dubé, M. 2003. Cumulative effect assessment in Canada: A regional framework for aquatic ecosystems. *Environmental Impact Assessment Review* 23:723–45.

Dubé, M., and K. Munkittrick. 2001. Integration of effects-based and stressor-based approaches into a holistic framework for cumulative effects assessment in aquatic ecosystems. *Human and Ecological Risk Assessment* 7:247–58.

Duchesne, M., S. D. Cote, and C. Barrette. 2000. Responses of woodland caribou to winter ecotourism in the Charlevoix Biosphere Reserve, Canada. *Biological Conservation* 96:311–7.

Dudgeon, D., A. H. Arthington, M. O. Gessner, Z. I. Kawabata, D. J. Knowler, C. Leveque, R. J. Naiman, et al. 2006. Freshwater biodiversity: Importance, threats, status and conservation challenges. *Biological Reviews* 81:163–82.

Duinker, P. N., and L. A. Greig. 2006. The importance of cumulative effects assessment in Canada: Ailments and ideas for redeployment. *Environmental Management* 37:153–61.

———. 2007. Scenario analysis in environmental impact assessment: Improving explorations of the future. *Environmental Impact Assessment Review* 27:206–19.

Dungan, J. L., J. N. Perry, M. R. T. Dale, P. Legendre, S. Citron-Pousty, M. J. Fortin, A. Jakomulska, M. Miriti, and M. S. Rosenberg. 2002. A balanced view of scale in spatial statistical analysis. *Ecography* 25:626–40.

Dussault, C., J. P. Ouellet, R. Courtois, J. Huot, L. Breton, and H. Jolicoeur. 2005. Linking moose habitat selection to limiting factors. *Ecography* 28:619–28.

Dyer, S. J., J. P. O'Neill, S. M. Wasel, and S. Boutin. 2001. Avoidance of industrial development by woodland caribou. *Journal of Wildlife Management* 65:531–42.

———. 2002. Quantifying barrier effects of roads and seismic lines on movements of female woodland caribou in northeastern Alberta. *Canadian Journal of Zoology* 80:839–45.

Easterly, T., A. Wood, and T. Litchfield. 1991. Responses of pronghorn and mule deer to petroleum development on crucial winter range in the Rattlesnake Hills. Completion Report A-1372. Cheyenne: Wyoming Game and Fish Department.

Ebeling, J., and M. Yasue. 2009. The effectiveness of market-based conservation in the tropics: Forest certification in Ecuador and Bolivia. *Journal of Environmental Management* 90:1145–53.

Eberhardt, L. L. 2002. A paradigm for population analysis of long-lived vertebrates. *Ecology* 83:2841–54.

Edge, W. 1982. *Distribution, habitat use, and movements of elk in relation to roads and human disturbances in western Montana*. M.S. thesis, University of Montana. Missoula.

Ehrenfeld, J. G. 2003. Effects of exotic plant invasions on soil nutrient cycling processes. *Ecosystems* 6:503–23.

Ehrlich, P. R., and R. M. Pringle. 2008. Where does biodiversity go from here? A grim business-as-usual forecast and a hopeful portfolio of partial solutions. *Proceedings of the National Academy of Sciences of the United States of America* 105:11579–86.

Ellis, K. L. 1984. Behavior of lekking sage grouse in response to a perched golden eagle. *Western Birds* 15:37–8.

Elton, C. 1958. *The ecology of invasions by plants and animals*. London: Methuen.

EnCana. 2007. *Environmental impact statement for the EnCana Shallow Gas Infill Development in the CFB Suffield National Wildlife Area*. Volume 1. www.ceaa.gc .ca/050/documents_staticpost/cearref_15620/131.pdf. Retrieved August 6, 2008.

Energy Information Administration. 2007. EIA International energy outlook. Report Number DOE/EIA 0484 (2007). Washington, DC: U.S. Department of Energy.

———. 2008. *Annual energy review 2008*. Washington, DC: U.S. Department of Energy.

———. 2009a. *Annual energy review 2009*. Washington, DC: U.S. Department of Energy.

———. 2009b. *Domestic uranium production report (2003–2008)*. Report Form EIA-851A. Washington, DC: U.S. Department of Energy.

Environmental Defense Fund. 1999. *Mitigation banking as an endangered species conservation tool*. Washington, DC: Environmental Defense Fund.

Environmental Law Institute. 2002. *Banks and fees: The status of off-site wetland mitigation in the United States*. Washington, DC: Environmental Law Institute.

———. 2007. *Mitigation of impacts to fish and wildlife habitat: Estimating costs and identifying opportunities*. Washington, DC: Environmental Law Institute.

Erickson, W. P., J. Jeffrey, K. Kronner, and K. Bay. 2004. *Stateline wind project wildlife monitoring final report July 2001–December 2003*. Technical report for and peer-reviewed by FPL Energy, Stateline Technical Advisory Committee, and the Oregon Energy Facility Siting Council, by Western EcoSystems Technology, Inc., Cheyenne, WY, and Northwest Wildlife Consultants, Pendleton, OR.

Erickson, W. P., J. Jeffrey, and V. K. Poulton. 2008. *Avian and bat monitoring: Year 1*. Puget Sound Energy Wild Horse Wind Project, Kittitas County, Washington. Prepared for Puget Sound Energy, Ellensburg, WA by Western EcoSystems Technology, Inc., Cheyenne, WY.

Erickson, W. P., G. D. Johnson, M. D. Strickland, and K. Kronner. 2000. *Avian and bat mortality associated with the Vansycle Wind Project, Umatilla County, Oregon: 1999 study year*. Technical report prepared for Umatilla County Department of Resource Services and Development by Western EcoSystems Technology, Inc., Cheyenne, WY.

Erickson, W. P., G. D. Johnson, M. D. Strickland, D. P. Young, K. J. Sernka, and R. E. Good. 2001. *Avian collisions with wind turbines: A summary of existing studies and comparisons to other sources of avian collision mortality in the United States*. Cheyenne, WY: Western EcoSystems Technology, Inc.

Erickson, W. P., K. Kronner, and B. Gritski. 2003. *Nine Canyon Wind Power Project avian and bat monitoring*. Prepared for Nine Canyon Technical Advisory Committee and Energy Northwest by Western EcoSystems Technology, Inc. Cheyenne, WY, and Northwest Wildlife Consultants, Pendleton, OR.

ESRI. 2006. *Terrestrial biomes in ESRI® Data and Maps*. Redlands, CA: ESRI.

Estes, J. A., E. M. Danner, D. F. Doak, B. Konar, A. M. Springer, P. D. Steinberg, M. T. Tinker, and T. M. Williams. 2004. Complex trophic interactions in kelp forest ecosystems. *Bulletin of Marine Science* 74:621–38.

Evangelista, P. H., S. Kumar, T. J. Stohlgren, C. S. Jarnevich, A. W. Crall, J. B. Norman, and D. T. Barnett. 2008. Modelling invasion for a habitat generalist and a specialist plant species. *Diversity and Distributions* 14:808–17.

Evangelista, P., T. J. Stohlgren, D. Guenther, and S. Stewart. 2004. Vegetation response to fire and postburn seeding treatments in juniper woodlands of the Grand Staircase–Escalante National Monument, Utah. *Western North American Naturalist* 64:293–305.

Evans, J. S., and S. A. Cushman. 2009. Gradient modeling of conifer species using random forests. *Landscape Ecology* 24:673–83.

Ewel, J., ed. 1986. *Invasibility: Lessons from south Florida*. New York: Springer.

Fahrig, L., and T. Rytwinski. 2009. Effects of roads on animal abundance: An empirical review and synthesis. *Ecology and Society* 14:21.

Fargione, J. E., T. R. Cooper, D. J. Flaspohler, J. Hill, C. Lehman, T. McCoy, S. McLeod, E. J. Nelson, K. S. Oberhauser, and D. Tilman. 2009. Bioenergy and wildlife: Threats and opportunities for grassland conservation. *BioScience* 59:767–77.

Fargione, J., J. Hill, D. Tilman, S. Polasky, and P. Hawthorne. 2008. Land clearing and the biofuel carbon debt. *Science* 319:1235–8.

Federal Advisory Committee. 2009. Wind Turbine Guidelines Advisory Committee synthesis workgroup draft version 4 for release to FAC August 9, 2009.

Federal Interagency Mitigation Workgroup. 2002. National wetlands mitigation action plan. December 24, 2002.

Fedy, B. C., K. Martin, C. Ritland, and J. Young. 2008. Genetic and ecological data provide incongruent interpretations of population structure and dispersal in naturally subdivided populations of white-tailed ptarmigan (*Lagopus leucura*). *Molecular Ecology* 17:1905–17.

Fenneman, N. M., and D. W. Johnson. 1946. *Physiographic divisions of the conterminous U.S* (map). water.usgs.gov/GIS/metadata/usgswrd/XML/physio.xml. Retrieved January 20, 2009.

Festa-Bianchet, M., J. M. Gaillard, and S. D. Cote. 2003. Variable age structure and apparent density dependence in survival of adult ungulates. *Journal of Animal Ecology* 72:640–9.

Findlay, C. S., S. Elgie, B. Giles, and L. Burr. 2009. Species listing under Canada's Species at Risk Act. *Conservation Biology* 23:1609–17.

Fischman, R. L. 2007. From words to action: The impact and legal status of the 2006 National Refuge System Management Policies. *Stanford Environmental Law Journal* 26:118–28.

Flaspohler, D. J., S. A. Temple, and R. N. Rosenfield. 2001. Species-specific edge effects on nest success and breeding bird density in a forested landscape. *Ecological Applications* 11:32–46.

Fleming, W. D., and F. K. A. Schmiegelow. 2003. *Response of bird communities to pipeline rights-of-way in the boreal forest of Alberta*. Proceedings of Seventh International Symposium on Environmental Concerns in Rights-of-Way Management.

Fletcher, Q. E., C. W. Dockrill, D. J. Saher, and C. L. Aldridge. 2003. Northern harrier, *Circus cyaneus*, attacks on greater sage-grouse, *Centerocercus urophasianus*, in southern Alberta. *Canadian Field-Naturalist* 117:479–80.

Folke, C., S. Carpenter, B. Walker, M. Scheffer, T. Elmqvist, L. Gunderson, and C. S. Holling. 2004. Regime shifts, resilience, and biodiversity in ecosystem management. *Annual Review of Ecology Evolution and Systematics* 35:557–81.

Forman, R. T. T., and L. E. Alexander. 1998. Roads and their major ecological effects. *Annual Review of Ecology and Systematics* 29:207–31.

Forman, R. T. T., B. Reineking, and A. M. Hersperger. 2002. Road traffic and nearby grassland bird patterns in a suburbanizing landscape. *Environmental Management* 29:782–800.

Fortin, D., H. L. Beyer, M. S. Boyce, D. W. Smith, T. Duchesne, and J. S. Mao. 2005. Wolves influence elk movements: Behavior shapes a trophic cascade in Yellowstone National Park. *Ecology* 86:1320–30.

Fortin, D., R. Courtois, P. Etcheverry, C. Dussault, and A. Gingras. 2008. Winter selection of landscapes by woodland caribou: Behavioural response to geographical gradients in habitat attributes. *Journal of Applied Ecology* 45:1392–400.

Fowler, G. S. 1999. Behavioral and hormonal responses of magellanic penguins (*Spheniscus magellanicus*) to tourism and nest site visitation. *Biological Conservation* 90:143–9.

Frair, J. L. 2005. *Survival and movement behaviour of resident and translocated wapiti (*Cervus elaphus*): Implications for their management in west-central Alberta, Canada*. Ph.D. dissertation, University of Alberta, Edmonton, Alberta, Canada.

Frair, J. L., E. H. Merrill, H. L. Beyer, and J. M. Morales. 2008. Thresholds in landscape connectivity and mortality risks in response to growing road networks. *Journal of Applied Ecology* 45:1504–13.

Frame, P. F., H. D. Cluff, and D. S. Hik. 2007. Response of wolves to experimental disturbance at homesites. *Journal of Wildlife Management* 71:316–20.

———. 2008. Wolf reproduction in response to caribou migration and industrial development on the central barrens of mainland Canada. *Arctic* 61:134–42.

Francis, C. D., C. P. Ortega, and A. Cruz. 2009. Noise pollution changes avian communities and species interactions. *Current Biology* 19:1415–9.

Fraser, S. J. 2007. Filling a public policy gap in Canada: Forest certification. *Forestry Chronicle* 83:666–71.

Frawley, B. J. 1989. *The dynamics of nongame bird breeding ecology in Iowa alfalfa fields*. M.S. thesis, Iowa State University, Ames.

Frayer, W. E., and G. M. Furnival. 1999. Forest survey sampling designs: A history. *Journal of Forestry* 97:4–10.

Freilich, J., B. Budd, T. Kohley, and B. Hayden. 2001. *The Wyoming basins ecoregional plan*. Lander: The Nature Conservancy, Wyoming Field Office.

Frid, A., and L. M. Dill. 2002. Human-caused disturbance stimuli as a form of predation risk. *Conservation Ecology* 6:11.

Fuller, T. K., L. D. Mech, and J. F. Cochrane, eds. 2003. *Wolf population dynamics*. Chicago: University of Chicago Press.

Gabrielsen, G. W., and E. N. Smith, eds. 1995. *Physiological responses of wildlife to disturbance*. Washington, DC: Island Press.

Gaillard, J. M., M. Festa-Bianchet, N. G. Yoccoz, A. Loison, and C. Toigo. 2000. Temporal variation in fitness components and population dynamics of large herbivores. *Annual Review of Ecology and Systematics* 31:367–93.

Galanti, V., D. Preatoni, A. Martinoti, L. A. Wauters, and G. Tosi. 2006. Space and habitat use of the African elephant in the Tarangire–Manyara ecosystem, Tanzania: Implications for conservation. *Mammalian Biology* 71:99–114.

Garton, E. O., J. W. Connelly, C. A. Hagen, J. S. Horne, A. Moser, and M. A. Schroeder. 2011. Greater sage-grouse population dynamics and probability of persistence. Number 12 in S. T. Knick and J. W. Connelly, eds. Greater sage-grouse: Ecology and conservation of a landscape species and its habitats. *Studies in Avian Biology*. sagemap.wr.usgs.gov/monograph.aspx. Retrieved September 10, 2010.

Gauthier, D. A., and E. B. Wiken. 2003. Monitoring the conservation of grassland habitats, Prairie Ecozone, Canada. *Environmental Monitoring and Assessment* 88:343–64.

Gaveau, D. L. A., M. Linkie, Suyadi, P. Levang, and N. Leader-Williams. 2009. Three decades of deforestation in southwest Sumatra: Effects of coffee prices, law enforcement and rural poverty. *Biological Conservation* 142:597–605.

Gelbard, J. L., and J. Belnap. 2003. Roads as conduits for exotic plant invasions in a semiarid landscape. *Conservation Biology* 17:420–32.

Gelbard, J. L., and S. Harrison. 2003. Roadless habitats as refuges for native grasslands: Interactions with soil, aspect, and grazing. *Ecological Applications* 13:404–15.

Gerhart, K. L., D. E. Russell, D. VanDewetering, R. G. White, and R. D. Cameron. 1997. Pregnancy of adult caribou (*Rangifer tarandus*): Evidence for lactational infertility. *Journal of Zoology* 242:17–30.

Gerrodette, T. 1987. A power analysis for detecting trends. *Ecology* 68:1364–72.

Gibbons, P., and D. B. Lindenmayer. 2007. Offsets for land clearing: No net loss or the tail wagging the dog? *Ecological Management and Restoration* 8:26–31.

Gibbs, J. P., ed. 2000. *Monitoring populations*. New York: Columbia University Press.

Gibbs, J. P., S. Droege, and P. Eagle. 1998. Monitoring populations of plants and animals. *BioScience* 48:935–40.

Gill, A. B. 2005. Offshore renewable energy: Ecological implications of generating electricity in the coastal zone. *Journal of Applied Ecology* 42:605–15.

Gill, J. P., M. Townsley, and G. P. Mudge. 1996. Review of the impacts of wind farms and other aerial structures upon birds. *Scottish Natural Heritage Review* 21.

Gill, R. B. 2001. *Declining mule deer populations in Colorado: Reasons and responses*. Report to the Colorado Legislature. Fort Collins: Colorado Division of Wildlife.

Gillies, C. S., M. Hebblewhite, S. E. Nielsen, M. A. Krawchuk, C. L. Aldridge, J. L. Frair, D. J. Saher, C. E. Stevens, and C. L. Jerde. 2006. Application of random effects to the study of resource selection by animals. *Journal of Animal Ecology* 75:887–98.

Gillin, C. M. 1989. *Response of elk to seismograph exploration in the Wyoming range, Wyoming*. M.S. thesis, University of Wyoming, Laramie.

Gillin, C. M., and L. Irwin. 1985. *Response of elk to seismograph exploration in the Bridger-Teton National Forest, Wyoming*. Laramie: University of Wyoming.

Girard, M., and B. Stotts. 1986. Managing impacts of oil and gas development on woodland wildlife habitats on the Little Missouri Grasslands, North Dakota. Pages 128–30 in *Proceedings II: Issues and technology in the management of impacted western wildlife*. Boulder, CO: Thorne Ecological Institute.

Gleason, R. A., and J. N. H. Euliss. 1998. Sedimentation of prairie wetlands. *Great Plains Research* 8:97–112.

Gordon, I. J., A. J. Hester, and M. Festa-Bianchet. 2004. The management of wild large herbivores to meet economic, conservation and environmental objectives. *Journal of Applied Ecology* 41:1021–31.

Government Accountability Office (GAO). 1998. Federal Advisory Committee

Act: Views of committee members and agencies on Federal Advisory Committee issues 5.

———. 2006. *GAO-06-730 Endangered species: Many factors affect the length of time to recover select species*. Washington, DC: U.S. Government Accountability Office.

———. 2007. *Agricultural conservation: Farm Program payments are an important factor in landowners' decisions to convert grassland to cropland*. GAO report GAO-07-1054. Washington, DC: U.S. Government Accountability Office.

Government of Canada. 2008. EnCana shallow gas infill development project-Canadian Environmental Assessment Registry Number 05-05-15620. Submission to the Canadian Environmental Assessment Agency.

Government of New South Wales. 2003. Native Vegetation Act 2003, Number 103.

Green, B. 1978. *Eating oil: Energy use in food production*. Boulder, CO: Westview.

Green, M. T., P. W. Lowther, S. L. Jones, S. K. Davis, and B. C. Dale. 2002. Baird's sparrow (*Ammodramus bairdii*). A. Poole, and F. Gill, eds. *The Birds of North America* Number 638. Philadelphia: The Birds of North America Inc.

Groves, C. R. 2003. *Drafting a conservation blueprint: A practitioner's guide to planning for biodiversity*. Washington, DC: Island Press.

Gucinski, H., M. J. Furniss, R. R. Ziemer, and M. H. Brookes, eds. 2001. *Forest roads: A synthesis of scientific information*. General Technical Report PNW-GTR-509. Portland, OR: Pacific Northwest Research Station, U.S. Forest Service.

Guisan, A., and N. E. Zimmermann. 2000. Predictive habitat distribution models in ecology. *Ecological Modelling* 135:147–86.

Gunderson, L., and C. Holling. 2002. *Panarchy: Understanding transformations in human and natural systems*. Washington, DC: Island Press.

Gustafson, E. J., D. E. Lytle, R. Swaty, and C. Loehle. 2007. Simulating the cumulative effects of multiple forest management strategies on landscape measures of forest sustainability. *Landscape Ecology* 22:141–56.

Gutzwiller, K. J., E. A. Kroese, S. H. Anderson, and C. A. Wilkins. 1997. Does human intrusion alter the seasonal timing of avian song during breeding periods? *Auk* 114:55–65.

Gutzwiller, K. J., R. T. Wiedenmann, K. L. Clements, and S. H. Anderson. 1994. Effects of human intrusion on song occurrence and singing consistency in subalpine birds. *Auk* 111:28–37.

Habib, L., E. M. Bayne, and S. Boutin. 2007. Chronic industrial noise affects pairing success and age structure of ovenbirds *Seiurus aurocapilla*. *Journal of Applied Ecology* 44:176–84.

Hagen, C. A. 2003. *A demographic analysis of lesser prairie-chicken populations in southwestern Kansas: Survival, population viability, and habitat use*. Ph.D. dissertation, Kansas State University, Manhattan.

Haight, R. G., B. Cypher, P. A. Kelly, S. Phillips, H. P. Possingham, K. Ralls, A. M. Starfield, P. J. White, and D. Williams. 2002. Optimizing habitat protection using demographic models of population viability. *Conservation Biology* 16:1386–97.

Hamilton, L. E. 2009. *Effects of natural gas development on three grassland bird species in CFB Suffield, Alberta, Canada.* M.S. thesis, University of Alberta, Edmonton.

Hannon, S. J., C. A. Paszkowski, S. Boutin, J. DeGroot, S. E. Macdonald, M. Wheatley, and B. R. Eaton. 2002. Abundance and species composition of amphibians, small mammals, and songbirds in riparian forest buffer strips of varying widths in the boreal mixedwood of Alberta. *Canadian Journal of Forest Research* 32:1784–1800.

Hanser, S. E., and S. T. Knick. 2011. Greater sage-grouse as an umbrella species for shrubland passerine birds: A multiscale assessment. Number 20 in S. T. Knick and J. W. Connelly, eds. Greater sage-grouse: Ecology and conservation of a landscape species and its habitats. *Studies in Avian Biology.* sagemap.wr.usgs .gov/monograph.aspx. Retrieved September 10, 2010.

Hansson, L. 1979. Field signs as indicators of vole abundance. *Journal of Applied Ecology* 16:339–47.

Hanusch, M., and J. Glasson. 2008. Much ado about SEA/SA monitoring: The performance of English Regional Spatial Strategies, and some German comparisons. *Environmental Impact Assessment Review* 28:601–17.

Hardner, J., and R. Rice. 2002. Rethinking green consumerism. *Scientific American* 286: 88–95.

Harju, S. M., M. R. Dzialak, R. C. Taylor, L. D. Hayden-Wing, and J. B. Winstead. 2010. Thresholds and time lags in effects of energy development on greater sage-grouse populations. *Journal of Wildlife Management* 74:437–48.

Harriman, J. A., and B. F. Noble. 2008. Characterizing project and strategic approaches to regional cumulative effects assessment in Canada. *Journal of Environmental Assessment Policy and Management* 10:25–50.

Harris, D. R. 1966. Recent plant invasions in arid and semi-arid Southwest of the United States. *Annals of the Association of American Geographers* 56:408–22.

Harrison, S., and E. Bruna. 1999. Habitat fragmentation and large-scale conservation: What do we know for sure? *Ecography* 22:225–32.

Harron, D. E. 2007. Assumptive error and overestimation of effects of in wildlife model output. *Ecology and Society* 12:3.

Harvey, B. C., and S. F. Railsback. 2007. Estimating multi-factor cumulative watershed effects on fish populations with an individual-based model. *Fisheries* 32:292–98.

Haskell, S. P., and W. B. Ballard. 2008. Annual re-habituation of calving caribou to oilfields in northern Alaska: Implications for expanding development. *Canadian Journal of Zoology* 86:627–37.

Haskell, S. P., R. M. Nielson, W. B. Ballard, M. A. Cronin, and T. L. McDonald. 2006. Dynamic responses of calving caribou to oilfields in northern Alaska. *Arctic* 59:179–90.

Haugrud, K. J. 2009. Perspectives on NEPA: Let's bring a bit of substance to NEPA-making mitigation mandatory. *Environmental Law Reporter* 39:10638–9.

Hayden-Wing Associates. 1990. Response of elk to Exxon's field development in the Riley Ridge Area of western Wyoming, 1979–1990. Final report prepared for Exxon Company, U.S.A. and Wyoming Game and Fish Department, Cheyenne, WY.

———. 1991. *Review and evaluation of the effects of geophysical exploration on some wildlife species in Wyoming*. Rawlins, WY: Geophysical Acquisition Workshop.

Hebblewhite, M., and E. Merrill. 2008. Modelling wildlife–human relationships for social species with mixed-effects resource selection models. *Journal of Applied Ecology* 45:834–44.

Hebblewhite, M., E. H. Merrill, and T. L. McDonald. 2005a. Spatial decomposition of predation risk using resource selection functions: An example in a wolf–elk predator–prey system. *Oikos* 111:101–11.

Hebblewhite, M., E. H. Merrill, L. E. Morgantini, C. A. White, J. R. Allen, E. Bruns, L. Thurston, and T. E. Hurd. 2006. Is the migratory behavior of montane elk herds in peril? The case of Alberta's Ya Ha Tinda elk herd. *Wildlife Society Bulletin* 34:1280–94.

Hebblewhite, M., C. A. White, C. G. Nietvelt, J. A. McKenzie, T. E. Hurd, J. M. Fryxell, S. E. Bayley, and P. C. Paquet. 2005b. Human activity mediates a trophic cascade caused by wolves. *Ecology* 86:2135–44.

Hettenhaus, J. 2004. *Biomass feedstock supply: As secure as the pipeline to a Naphtha Cracker?* Presentation at World Congress on industrial biotechnology and bioprocessing. Orlando, Florida.

Hiatt, G. S., and D. Baker. 1981. *Effects of oil/gas drilling on elk and mule deer winter distributions on Crooks Mountain, Wyoming*. Cheyenne: Wyoming Game and Fish Department.

Higgins, K. F., R. G. Osborn, and D. E. Naugle. 2007. Effects of wind turbines on birds and bats in southwestern Minnesota, U.S.A. Pages 153–75 in M. de Lucas G. F. E. Janss, and M. Ferrer, eds. *Birds and wind farms: Risk assessment and mitigation*. Madrid: Quercus Press.

Hill, D. P., and L. K. Gould. 1997. Chestnut-collared longspur (*Calcarius ornatus*). A. Poole, and F. Gill, eds. *The Birds of North America* Number 288. Philadelphia: The Birds of North America, Inc.

Hins, C., J. P. Ouellet, C. Dussault, and M. H. St-Laurent. 2009. Habitat selection by forest-dwelling caribou in managed boreal forest of eastern Canada: Evidence of a landscape configuration effect. *Forest Ecology and Management* 257:636–43.

Hinz, H., V. Prieto, and M. J. Kaiser. 2009. Trawl disturbance on benthic communities: Chronic effects and experimental predictions. *Ecological Applications* 19:761–773.

Hobson, K. A., and J. Schieck. 1999. Changes in bird communities in boreal mixedwood forest: Harvest and wildfire effects over 30 years. *Ecological Applications* 9:849–63.

Hoekstra, J. M., T. M. Boucher, T. H. Ricketts, and C. Roberts. 2005. Confronting a biome crisis: Global disparities of habitat loss and protection. *Ecology Letters* 8:23–9.

Holling, C. S. 1973. Resilience and stability of ecological systems. *Annual Review of Ecology and Systematics* 4:1–23.

Holling, C. S., ed. 1978. *Adaptive environmental assessment and management*. New York: Wiley and Sons.

Holloran, M. J. 2005. *Greater sage-grouse (*Centrocercus urophasianus*) population response to gas field development in western Wyoming*. Ph.D. dissertation, University of Wyoming, Laramie.

Holloran, M. J., R. C. Kaiser, and W. A. Hubert. 2007. *Population response of yearling greater sage-grouse to the infrastructure of natural gas fields in southwestern Wyoming*. Completion report. Laramie, WY: U.S. Department of Interior, Geological Survey.

———. 2010. Yearling greater sage-grouse response to energy development in Wyoming. *Journal of Wildlife Management* 74:65–72.

Holloran, M. R. J., and S. H. Anderson. 2005. Spatial distribution of greater sage-grouse nests in relatively contiguous sagebrush habitats. *Condor* 107:742–52.

Hood, G. A., and K. L. Parker. 2001. Impact of human activities on grizzly bear habitat in Jasper National Park. *Wildlife Society Bulletin* 29:624–38.

Horesji, B. 1979. *Seismic operations and their impacts on large mammals: Results of a monitoring program*. Western Wildlife Environments Consulting, Calgary, Alberta, Canada.

Horton, J. S. 1977. *The development and perpetuation of the permanent Tamarisk type in the phreatophyte zone of the Southwest*. General Technical Report RM43. Washington, DC: U.S. Department of Agriculture, Forest Service.

Horwich, R. H., and J. Lyon. 2007. Community conservation: Practitioners' answer to critics. *Oryx* 41:376–85.

Houck, O. 2009a. Nature or nurture: What's wrong and what's right with adaptive management. *Environmental Law Reporter* 39:10923–4.

———. 2009b. How'd we get divorced?: The curious case of NEPA and planning. *Environmental Law Reporter* 39:10645–50.

Howell, J. A., and J. Noone. 1992. *Examination of avian use and mortality at a U.S. Windpower wind energy development site, Montezuma Hills, Solano County, California*. Final report to Solano County Department of Environmental Management.

Huenneke, L. F., S. P. Hamburg, R. Koide, H. A. Mooney, and P. M. Vitousek. 1990. Effects of soil resources on plant invasion and community structure in California serpentine grassland. *Ecology* 71:478–91.

Hulse, D., A. Branscomb, C. Enright, and J. Bolte. 2009. Anticipating floodplain trajectories: A comparison of two alternative futures approaches. *Landscape Ecology* 24:1067–90.

Hunt, W. G. 2002. Golden eagles in a perilous landscape: Predicting the effects of mitigation for wind turbine bladestrike mortality. California Energy Commission (CEC) Consultant Report P500-02-043F, CEC Sacramento, CA. Prepared for CEC, Public Interest Energy Research, Sacramento, CA, by University of California–Santa Cruz.

Hunt, G., and T. Hunt. 2006. *The trend of golden eagle territory occupancy in the vicinity of the Altamont Pass Wind Resource Area: 2005 survey*. Report CEC-500-2006-056. Sacramento, CA: Public Interest Energy Research Program.

Hurlbert, S. H. 1984. Pseudoreplication and the design of ecological field experiments. *Ecological Monographs* 54:187–211.

Ihsle, H. B. 1982. *Population ecology of mule deer with emphasis on potential impacts of gas and oil development along the east slope of the Rocky Mountains, north-central Montana*. M.S. thesis, Montana State University, Bozeman.

Ingelfinger, F., and S. Anderson. 2004. Passerine response to roads associated with natural gas extraction in a sagebrush steppe habitat. *Western North American Naturalist* 64:385–95.

International Energy Agency. 2007. *World energy outlook 2007*. Paris: IEA.

———. 2008. *World energy outlook 2008*. Paris: IEA.

Irby, L. R., R. J. Mackie, H. I. Pac, and W. F. Kasworm. 1988. Management of mule deer in relation to oil and gas development in Montana's overthrust belt. Pages 113–121 in *Proceedings III: Issues and technology in the management of impacted wildlife*. Boulder, CO: Thorne Ecological Institute.

Irvine, J. R., and N. E. West. 1979. Riparian tree species distribution and succession along the Lower Escalante River, Utah. *Southwestern Naturalist* 24:331–46.

Irwin, L., and C. Gillian, eds. 1984. *Response of elk to seismic exploration in the Bridger-Teton Forest, Wyoming*. Laramie: University of Wyoming.

Ivanov, V. L. 1985. Prediction of the oil and gas potential of Antarctica on the basis of geological conditions. *Polar Geography* 9:116–31.

Jackson, T., and J. Curry. 2002. Regional development and land use planning in rural British Columbia: Peace in the woods? *Regional Studies* 36:439–43.

James, A. R. C., S. Boutin, D. M. Hebert, and A. B. Rippin. 2004. Spatial separation of caribou from moose and its relation to predation by wolves. *Journal of Wildlife Management* 68:799–809.

James, A. R. C., and A. K. Stuart-Smith. 2000. Distribution of caribou and wolves in relation to linear corridors. *Journal of Wildlife Management* 64:154–9.

Johnson, C. J. 2011. *Regulating and planning for cumulative effects: The Canadian experience. Cumulative effects in wildlife management: Impact mitigation*. New York: Taylor and Francis.

Johnson, K. 2006. GIS habitat analysis for lesser prairie-chickens in southeastern New Mexico. *BMC Ecology* 6:18.

Johnson, G. D. 2005. A review of bat mortality at wind-energy developments in the United States. *Bat Research News* 46:45–9.

Johnson, B. K., and D. Lockman. 1979. *Response of elk during calving to oil/gas drilling activity in Snider Basin, Wyoming*. Cheyenne: Wyoming Game and Fish Report, Wyoming Game and Fish Department.

Johnson, C. J., and M. S. Boyce. 2004. A quantitative approach for regional environmental assessment: Application of a habitat-based population viability analysis to wildlife of the Canadian central Arctic. Canadian Environmental Assessment Agency Research and Development Monograph Series.

Johnson, B., and L. Wollrab. 1987. *Response of elk to development of a natural gas field in western Wyoming 1979–1987*. Cheyenne: Riley Ridge Natural Gas Project, Wyoming Game and Fish Department.

Johnson, C. J., M. S. Boyce, R. L. Case, H. D. Cluff, R. J. Gau, A. Gunn, and R. Mulders. 2005. Cumulative effects of human developments on arctic wildlife. *Wildlife Monographs* 160.

Johnson, C. J., and M. P. Gillingham. 2004. Mapping uncertainty: Sensitivity of wildlife habitat ratings to expert opinion. *Journal of Applied Ecology* 41:1032–41.

Johnson, C. J., K. L. Parker, D. C. Heard, and M. P. Gillingham. 2002. Movement parameters of ungulates and scale-specific responses to the environment. *Journal of Animal Ecology* 71:225–35.

Johnson, D. H., and M. D. Schwartz. 1993. The Conservation Reserve Program and grassland birds. *Conservation Biology* 7:934–7.

Johnson, D. H., M. J. Holloran, J. W. Connelly, S. E Hanser, C. L. Amundson, and S. T. Knick. 2011. Influences of environmental and anthropogenic features on greater sage-grouse populations, 1997–2007. Number 17 in S. T. Knick and J. W. Connelly, eds. Greater sage-grouse: Ecology and conservation of a landscape species and its habitats. *Studies in Avian Biology*. sagemap.wr.usgs.gov/monograph.aspx. Retrieved September 10, 2010.

Johnson, E. A., and K. Miyanishi. 2008. Creating new landscapes and ecosystems: The Alberta oil sands. *Annals of the New York Academy of Sciences* 1134:120–45.

Johnson, G. D., and W. P. Erickson. 2008. *Avian and bat cumulative impacts associated with wind energy development in the Columbia Plateau ecoregion of eastern Washington and Oregon*. Cheyenne, WY: Western EcoSystems Technology, Inc.

Johnson, G. D., W. P. Erickson, M. D. Strickland, M. F. Shepherd, and D. A. Shepherd. 2000a. *Avian monitoring studies at the Buffalo Ridge Wind Resource Area, Minnesota: Results of a 4-year study*. Final report. Cheyenne, WY: Western EcoSystems Technology, Inc.

Johnson, G. D., W. P. Erickson, and J. White. 2003. *Avian and bat mortality during the first year of operation at the Klondike Phase I Wind Project, Sherman County, Oregon*. Technical report prepared for Northwestern Wind Power, Goldendale, WA by Western EcoSystems Technology, Inc., Cheyenne, WY.

Johnson, G. D., W. Erickson, and E. Young. 2009a. *Greater prairie chicken lek surveys, Elk River Wind Farm, Butler County, Kansas.* Unpublished report prepared for Iberdrola Renewables by Western EcoSystems Technology, Inc., Cheyenne, WY.

Johnson, G. D., C. LeBeau, T. Rintz, J. Eddy, and M. Holloran. 2009b. *Greater sage-grouse telemetry study for the Simpson Ridge Wind Energy Project, Carbon County, Wyoming: Second quarterly report.* Cheyenne, WY: Western EcoSystems Technology, Inc.

Johnson, R. R., C. K. Baxter, and M. E. Estey, eds. 2009c. An emerging agency-based approach to conserving populations through strategic habitat conservation. Pages 201–224 in J. J. Millspaugh and F. R. Thompson III, eds. *Models for planning wildlife conservation in large landscapes.* Burlington, MA: Academic Press.

Johnson, G. D., M. D. Strickland, W. P. Erickson, and J. D. P. Young, eds. 2007. Use of data to develop mitigation measures for wind power development impacts to birds. Pages 241–258 in M. de Lucas G. F. E. Janss, and M. Ferrer, eds. *Birds and wind farms: Risk assessment and mitigation.* Madrid: Quercus.

Johnson, G. D., D. P. Young, W. P. Erickson, C. E. Derby, M. D. Strickland, and R. E. Good. 2000b. *Wildlife monitoring studies, SeaWest Windpower Plant, Carbon County, Wyoming, 1995–1999.* Final report. Cheyenne, WY: Western Eco-Systems Technology, Inc.

Johnson, K., T. B. Neville, and P. Neville. 2006. GIS habitat analysis for lesser prairie-chickens in southeastern New Mexico. *BMC Ecology* 6:18.

Joly, K., C. Nellemann, and I. Vistnes. 2006. A reevaluation of caribou distribution near an oilfield road on Alaska's North Slope. *Wildlife Society Bulletin* 34:866–9.

Jones, J. P. G., M. M. Andriamarovololona, and N. Hockley. 2008. The importance of taboos and social norms to conservation in Madagascar. *Conservation Biology* 22:976–86.

Jones, S. L., and J. E. Cornely. 2002. Vesper sparrow (*Pooecetes gramineus*). A. Poole, and F. Gill, eds. *The Birds of North America* Number 624. Philadelphia: The Birds of North America, Inc.

Kaiser, R. C. 2006. *Recruitment by greater sage-grouse in association with natural gas development in western Wyoming.* M.S. thesis, University of Wyoming, Laramie.

Karkkainen, B. C. 2002. Toward a smarter NEPA: Monitoring and managing government's environmental performance. *Columbia Law Review* 102:903–72.

Keane, A., J. P. G. Jones, G. Edwards-Jones, and E. J. Milner-Gulland. 2008. The sleeping policeman: Understanding issues of enforcement and compliance in conservation. *Animal Conservation* 11:75–82.

Kemper, J. T., and S. E. Macdonald. 2009a. Directional change in upland tundra plant communities 20–30 years after seismic exploration in the Canadian low-arctic. *Journal of Vegetation Science* 20:557–67.

———. 2009b. Effects of contemporary winter seismic exploration on low arctic plant communities and permafrost. *Arctic Antarctic and Alpine Research* 41:228–37.

Kennedy, A. J. 2000. *Cumulative environmental effects management: Tools and approaches*. Calgary, AB: Hignell Printing.

Kennett, S. A. 1999. Towards a new paradigm for cumulative effects management. Canadian Institute of Resources Law CIRL Occasional Paper #8. University of Calgary, Edmonton, Alberta.

Kerbiriou, C., I. Le Viol, A. Robert, E. Porcher, F. Gourmelon, and R. Julliard. 2009. Tourism in protected areas can threaten wild populations: From individual response to population viability of the chough *Pyrrhocorax pyrrhocorax*. *Journal of Applied Ecology* 46:657–65.

Kerlinger, P., R. Curry, L. Culp, A. Jain, C. Wilkerson, B. Fischer, and A. Hasch. 2006. *Post-construction avian and bat fatality monitoring for the High Winds Wind Power Project, Solano County, California: Two year report*. Prepared for High Winds, LLC, FPL Energy by Curry and Kerlinger, LLC.

Kerr, J. T., and D. J. Currie. 1995. Effects of human activity on global extinction risk. *Conservation Biology* 9:1528–38.

Kiesecker, J. M., L. K. Belden, K. Shea, and M. J. Rubbo. 2004. Amphibian decline and emerging disease. *American Scientist* 92:138–47.

Kiesecker, J. M., H. Copeland, A. Pocewicz, and B. McKenney. 2010. Development by design: Blending landscape-level planning with the mitigation hierarchy. *Frontiers in Ecology and the Environment* 8:261–66. doi:10.1890/090005.

Kiesecker, J. M., H. Copeland, A. Pocewicz, N. Nibbelink, B. McKenney, J. Dahlke, M. Holloran, and D. Stroud. 2009. A framework for implementing biodiversity offsets: Selecting sites and determining scale. *BioScience* 59:77–84.

Kimerling, J. 2001. Corporate ethics in the era of globalization: The promise and peril of international environmental standards. *Journal of Agricultural and Environmental Ethics* 14:425–55.

King, D. M., and E. W. Price. 2004. *Developing defensible wetland mitigation ratios: A companion to the five-step wetland mitigation ratio calculator*. Solomons: Center for Environmental Science, University of Maryland.

Kiritani, K., and K. Yamamura, eds. 2003. *Exotic insects and their pathways for invasion*. Washington, DC: Island Press.

Kirk, R. E. 1996. Practical significance: A concept whose time has come. *Educational and Psychological Measurement* 56:746–59.

Knick, S. T., D. S. Dobkin, J. T. Rotenberry, M. A. Schroeder, W. M. Vander Haegen, and C. van Riper. 2003. Teetering on the edge or too late? Conservation and research issues for avifauna of sagebrush habitats. *Condor* 105:611–34.

Knick, S. T., and S. E. Hanser. 2011. Connecting pattern and process in greater sage-grouse populations and sagebrush landscapes. Number 18 in S. T. Knick and J. W. Connelly, eds. Greater sage-grouse: Ecology and conservation of a

landscape species and its habitats. *Studies in Avian Biology*. sagemap.wr.usgs
.gov/monograph.aspx. Retrieved September 10, 2010.

Knick, S. T., and J. T. Rotenberry. 1995. Landscape characteristics of fragmented
shrubsteppe habitats and breeding passerine birds. *Conservation Biology*
9:1059–71.

———. 1999. Spatial distribution of breeding passerine bird habitats in a shrub-
steppe region of southwestern Idaho. *Studies in Avian Biology* 19:104–11.

Knight, J. 1981. Effects of oil and gas development on elk movements and distribution in northern Michigan. *Transactions of the Forty-Sixth North American Wildlife and Natural Resources Conference* 46:349–57.

Knight, J. E., Jr. 1980. *Effects of hydrocarbon development on elk movements and distribution in northern Michigan*. Ph.D. dissertation, University of Michigan, Ann
Arbor.

Knight, R. L., and S. F. Bates. 1995. *A new century for natural resources management*.
Washington, DC: Island Press.

Kofinas, G., P. Lyver, D. Russell, R. White, A. Nelson, and N. Flanders. 2003. Towards a protocol for community monitoring of caribou body condition.
Rangifer 14:43–52.

Köller, J., J. Köppel, and W. Peters, eds. 2006. *Offshore wind energy: Research on environmental impacts*. New York: Springer.

Kotliar, N. B., and J. A. Wiens. 1990. Multiple scales of patchiness and patch structure: A hierarchical framework for the study of heterogeneity. *Oikos* 59:253–
60.

Krausman, P. R., L. K. Harris, C. L. Blasch, K. K. G. Koenen, and J. Francine. 2004.
Effects of military operations on behavior and hearing of endangered Sonoran
pronghorn. *Wildlife Monographs* 157.

Krebs, C. J. 1989. *Ecological methodology*. New York: HarperCollins.

———. 1999. *Ecological methodology*, 2nd ed. Don Mills, Ontario: Addison-Wesley
Educational Publishers.

Krebs, J. R., and N. B. Davies. 1993. *An introduction to behavioural ecology*, 3rd ed.
Oxford: Blackwell Science.

Kronner, K., B. Gritski, and S. Downes. 2008. *Big Horn Wind Power Project wildlife
fatality monitoring study: 2006–2007*. Final report prepared for PPM Energy
and the Big Horn Wind Project Technical Advisory Committee by Northwest
Wildlife Consultants, Inc., Goldendale, WA.

Kuck, L., G. L. Hompland, and E. H. Merrill. 1985. Elk calf response to simulated mine disturbance in southeast Idaho. *Journal of Wildlife Management*
49:751–7.

Kucks, R. P., and P. I. Hill. 2000. Wyoming aeromagnetic and gravity maps and
data: A Web site for distribution of data. Open file report 00-0198 crustal.usgs
.gov/geophysics/state.html. Retrieved August 30, 2009.

Kumar, C. 2005. Revisiting "community" in community-based natural resource
management. *Community Development Journal* 40:275–85.

Kunkel, K. E., and D. H. Pletscher. 2000. Habitat factors affecting vulnerability of moose to predation by wolves in southeastern British Columbia. *Canadian Journal of Zoology* 78:150–7.

Kunz, T. H., E. B. Arnett, B. M. Cooper, W. P. Erickson, R. P. Larkin, T. Mabee, M. L. Morrison, M. D. Strickland, and J. M. Szewczak. 2007a. Assessing impacts of wind-energy development on nocturnally active birds and bats: A guidance document. *Journal of Wildlife Management* 71:2449–86.

Kunz, T. H., E. B. Arnett, W. P. Erickson, A. R. Hoar, G. D. Johnson, R. P. Larkin, M. D. Strickland, R. W. Thresher, and M. D. Tuttle. 2007b. Ecological impacts of wind energy development on bats: Questions, research needs, and hypotheses. *Frontiers in Ecology and the Environment* 5:315–24.

Kusler, J. 2003. *Recommendations for reconciling wetland assessment techniques*. Association of State Wetland Managers. Institute for Wetland Science and Public Policy, Berne, New York.

Kuvlesky, W. P., L. A. Brennan, M. L. Morrison, K. K. Boydston, B. M. Ballard, and F. C. Bryant. 2007. Wind energy development and wildlife conservation: Challenges and opportunities. *Journal of Wildlife Management* 71:2487–98.

LaGory, K. E., Y. Chang, K. C. Chun, T. Reeves, R. Liebich, and K. Smith. 2001. A study of the effects of gas well compressor noise on breeding bird populations of the Rattlesnake Canyon Habitat Management Area, San Juan County, New Mexico. Report DOE/BC/W-31-109-ENG-38-10. Argonne, IL: Argonne National Laboratory.

Laliberte, A. S., and W. J. Ripple. 2003. Wildlife encounters by Lewis and Clark: A spatial analysis of interactions between native Americans and wildlife. *BioScience* 53:994–1003.

Laliberte, A. S., and W. J. Ripple. 2004. Range contractions of North American carnivores and ungulates. *BioScience* 54:123–38.

Lancaster, P. A., J. Bowman, and B. A. Pond. 2008. Fishers, farms, and forests in eastern North America. *Environmental Management* 42:93–101.

Lancia, R. A., J. D. Nichols, and K. H. Pollock, eds. 1994. *Estimating the number of animals in wildlife populations*, 5th ed. Bethesda, MD: The Wildlife Society.

Larson, D. L., P. J. Anderson, and W. Newton. 2001. Alien plant invasion in mixed-grass prairie: Effects of vegetation type and anthropogenic disturbance. *Ecological Applications* 11:128–41.

Laurian, C., C. Dussault, J. P. Ouellet, R. Courtois, M. Poulin, and L. Breton. 2008. Behavior of moose relative to a road network. *Journal of Wildlife Management* 72:1550–7.

Lawrence, D. P. 2003. *Environmental impact assessment: Practical solutions to recurrent problems*. Hoboken, NJ: Wiley and Sons.

Lawton, J. H., D. E. Bignell, B. Bolton, G. F. Bloemers, P. Eggleton, P. M. Hammond, M. Hodda, et al. 1998. Biodiversity inventories, indicator taxa and effects of habitat modification in tropical forest. *Nature* 391:72–6.

Leddy, K. L. 1996. *Effects of wind turbines on nongame birds in Conservation Reserve Program grasslands in southwestern Minnesota*. M.S. thesis, South Dakota State University, Brookings.

Leddy, K. L., K. F. Higgins, and D. E. Naugle. 1999. Effects of wind turbines on upland nesting birds in Conservation Reserve Program grasslands. *Wilson Bulletin* 111:100–4.

Lee, C. E. 2002. Evolutionary genetics of invasive species. *Trends in Ecology and Evolution* 17:386–91.

Lee, K. N. 1993. *Compass and gyroscope: Integrating science and politics for the environment*. Washington, DC: Island Press.

Lee, P., and S. Boutin. 2006. Persistence and developmental transition of wide seismic lines in the western Boreal Plains of Canada. *Journal of Environmental Management* 78:240–50.

Lessard, R. B., S. J. D. Martell, C. J. Walters, T. E. Essington, and J. F. Kitchell. 2005. Should ecosystem management involve active control of species abundances? *Ecology and Society* 10:1.

Leu, M., S. E. Hanser, and S. T. Knick. 2008. The human footprint in the west: A large-scale analysis of anthropogenic impacts. *Ecological Applications* 18:1119–39.

Levin, P. S., and D. A. Levin. 2004. The real biodiversity crisis. *American Scientist* 90:6–8.

Levin, S. A. 1992. The problem of pattern and scale in ecology. *Ecology* 73:1943–67.

Levine, J. M., and C. M. D'Antonio. 1999. Elton revisited: A review of evidence linking diversity and invasibility. *Oikos* 87:15–26.

Levine, J. M., M. Vila, C. M. D'Antonio, J. S. Dukes, K. Grigulis, and S. Lavorel. 2003. Mechanisms underlying the impacts of exotic plant invasions. *Proceedings of the Royal Society of London Series B: Biological Sciences* 270:775–81.

Linnell, J. D. C., R. Aanes, and R. Andersen. 1995. Who killed Bambi? The role of predation in the neonatal mortality of temperature ungulates. *Wildlife Biology* 1:209–23.

Linnen, C. G. 2006. *The effects of minimal disturbance shallow gas activity on grassland birds*. Northern EnviroSearch Limited. Prepared for the Petroleum Technology Alliance Canada.

———. 2008. *Effects of oil and gas development on grassland birds*. Northern EnviroSearch Limited. Prepared for Petroleum Technology Alliance Canada.

Lloyd, J. D., and T. E. Martin. 2005. Reproductive success of chestnut-collared longspurs in native and exotic grassland. *Condor* 107:363–74.

Lloyd, J. D., G. L. Slater, and S. Snow. 2009. Demography of reintroduced eastern bluebirds and brown-headed nuthatches. *Journal of Wildlife Management* 73:955–64.

Lodge, D. M. 1993. Biological invasions: Lessons for ecology. *Trends in Ecology and Evolution* 8:133–7.

Lodge, D. M., S. Williams, H. J. MacIsaac, K. R. Hayes, B. Leung, S. Reichard, R. N. Mack, et al. 2006. Biological invasions: Recommendations for U.S. policy and management. *Ecological Applications* 16:2035–54.

Lonsdale, W. M., and A. M. Lane. 1994. Tourist vehicles as vectors of weed seeds in Kakadu National Park, northern Australia. *Biological Conservation* 69:277–83.

Lovejoy, T. E. 1980. Discontinuous wilderness: Minimum areas for conservation. *Parks* 5:13–5.

Low, G. 2003. *Landscape-scale conservation: A practitioner's guide*. Arlington, VA: The Nature Conservancy.

Lozon, J. D., and H. J. MacIsaac. 1997. Biological invasions: Are they dependent on disturbance? *Environmental Reviews* 5:131–44.

Ludwig, D., M. Mangel, and B. Haddad. 2001. Ecology, conservation, and public policy. *Annual Review of Ecology and Systematics* 32:481–517.

Lustig, T. 2002. Where would you like the holes drilled into your crucial winter range? *Transactions of the North American Wildlife and Natural Resources Conference* 67:317–25.

Lyon, A. G., and S. H. Anderson. 2003. Potential gas development impacts on sage grouse nest initiation and movement. *Wildlife Society Bulletin* 31:486–91.

Mabey, S., and E. Paul. 2007. *Impact of wind energy and related human activities on grassland and shrub-steppe birds*. Literature review for National Wind Coordinating Collaborative and The Ornithological Council.

MacArthur, R. A., R. H. Johnston, and V. Geist. 1979. Factors influencing heart rate in free-ranging bighorn sheep: A physiological approach to the study of wildlife harassment. *Canadian Journal of Zoology* 57:2010–21.

MacFarlane, A. 2003. *Vegetation response to seismic lines: Edge effects and on-line succession*. M.S. thesis, University of Alberta, Edmonton.

Machtans, C. S. 2006. Songbird response to seismic lines in the western boreal forest: A manipulative experiment. *Canadian Journal of Zoology* 84:1421–30.

Mack, R. N., ed. 1986. *Alien plant invasion into the intermountain West: A case history*. New York: Springer.

Mack, R. N., and W. M. Lonsdale. 2001. Humans as global plant dispersers: Getting more than we bargained for. *BioScience* 51:95–102.

Mack, R. N., D. Simberloff, W. M. Lonsdale, H. Evans, M. Clout, and F. A. Bazzaz. 2000. Biotic invasions: Causes, epidemiology, global consequences, and control. *Ecological Applications* 10:689–710.

Maddox, D. M., A. Mayfield, and N. H. Poritz. 1985. Distribution of yellow starthistle (*Centaurea solstitialis*) and Russian knapweed (*Centaurea repens*). *Weed Science* 33:315–27.

Madsen, J. 1994. Impacts of disturbance on migratory waterfowl. *Ibis* 137:67–74.

———. 1998. Experimental refuges for migratory waterfowl in Danish wetlands. II. Tests of hunting disturbance effects. *Journal of Applied Ecology* 35:398–417.

Madsen, P. T., M. Wahlberg, J. Tougaard, K. Lucke, and P. Tyack. 2006. Wind turbine underwater noise and marine mammals: Implications of current knowledge and data needs. *Marine Ecology Progress Series* 309:279–95.

Mandelker, D. 2009. Thoughts on NEPA at 40. *Environmental Law Reporter* 39:10640–1.

Manes, R., S. Harmon, B. Obermeyer, and R. Applegate. 2002. *Wind energy and wildlife: An attempt at pragmatism.* Special report of the Wildlife Management Institute.

Margoluis, R. A., N. Salafsky, and A. Balla. 1998. *Measures of success: Designing, managing, and monitoring conservation and development projects.* Washington, DC: Island Press.

Martin, B., A. Pearson, and B. Bauer. 2009. *An ecological risk assessment of wind energy development in Montana.* Helena, MT: The Nature Conservancy.

Martin, J. W., and B. A. Carlson, eds. 1998. Sage sparrow (*Amphispiza belli*). A. Poole, and F. Gill, eds. *The Birds of North America* Number 326. Philadelphia: The Birds of North America, Inc.

Martin, P. S., and C. R. Szuter. 1999. War zones and game sinks in Lewis and Clark's West. *Conservation Biology* 13:36–45.

Massachusetts Institute of Technology. 2006. The future of geothermal energy: Impact of enhanced geothermal systems (EGS) on the United States in the 21st century. Idaho National Laboratories, Idaho Falls, Idaho.

Mattson, D. J., and T. Merrill. 2002. Extirpations of grizzly bears in the contiguous United States, 1850–2000. *Conservation Biology* 16:1123–36.

McCutchen, N. A. 2007. *Factors affecting caribou survival in northern Alberta: The role of wolves, moose, and linear features.* Ph.D. dissertation, University of Alberta, Edmonton.

McDonald, R. I., J. Fargione, J. Kiesecker, W. M. Miller, and J. Powell. 2009. Energy sprawl or energy efficiency: Climate policy impacts on natural habitat for the United States of America. *PLoS One* 4:e6802.

McGowan, C. P., and T. R. Simons. 2006. Effects of human recreation on the incubation behavior of American oystercatchers. *Wilson Journal of Ornithology* 118:485–93.

McKenney, B. A., and J. M. Kiesecker. 2010. Policy development for biodiversity offsets: A review of offset frameworks. *Environmental Management* 45:165–76.

McKenzie, H. W., M. A. Lewis, and E. H. Merrill. 2009. First passage time analysis of animal movement and insights into the functional response. *Bulletin of Mathematical Biology* 71:107–29.

McKinney, M. L. 2002. Do human activities raise species richness? Contrasting patterns in United States plants and fishes. *Global Ecology and Biogeography* 11:343–8.

McLoughlin, P. D., J. S. Dunford, and S. Boutin. 2005. Relating predation mortality to broad-scale habitat selection. *Journal of Animal Ecology* 74:701–7.

McLoughlin, P. D., E. Dzus, B. Wynes, and S. Boutin. 2003. Declines in populations of woodland caribou. *Journal of Wildlife Management* 67:755–61.

McMaster, D. G., and S. K. Davis. 2001. An evaluation of Canada's Permanent Cover Program: Habitat for grassland birds? *Journal of Field Ornithology* 72:195–210.

MEDIAS-France/Postel. 2004. North America Globcover V2.2. ESA/ESA Globcover Project.

Meffe, G. K., and S. Viederman. 1995. Combining science and policy in conservation biology. *Wildlife Society Bulletin* 23:327–32.

Melo, C. J., and S. A. Wolf. 2005. Empirical assessment of eco-certification: The case of Ecuadorian bananas. *Organization Environment* 18:287–317.

Mengel, R. M. 1970. The North American Central Plains as an isolating agent in bird speciation. Pages 279–340 in W. Dort, Jr., and J. K. Jones, Jr., eds. *Pleistocene and recent environments of the Central Great Plains*. Department of Geology Special Publication 3. Lawrence, Kansas: University Press of Kansas.

Merrill, E. H., T. P. Hemker, K. P. Woodruff, and L. Kuck. 1994. Impacts of mining facilities on fall migration of mule deer. *Wildlife Society Bulletin* 22:68–73.

Miller, S. G., R. L. Knight, and C. K. Miller. 1998. Influence of recreational trails on breeding bird communities. *Ecological Applications* 8:162–9.

Mills, L. S. 2007. Bridging applied population and ecosystem ecology with focal species concepts. Pages 276–285 in *Conservation of wildlife populations: Demography, genetics, and management*. Malden, MA: Blackwell.

Mills, L. S., D. F. Doak, and M. J. Wisdom. 1999. Reliability of conservation actions based on elasticity analysis of matrix models. *Conservation Biology* 13:815–29.

Mills, T. J., and R. N. Clark. 2001. Roles of research scientists in natural resource decision-making. *Forest Ecology and Management* 153:189–98.

Millspaugh, J. J., G. C. Brundige, R. A. Gitzen, and K. J. Raedeke. 2000. Elk and hunter space-use sharing in South Dakota. *Journal of Wildlife Management* 64:994–1003.

Montana Board of Oil and Gas Conservation. 2006. *Annual review 2006*, Vol. 50. Helena: Department of Natural Resources and Conservation, State of Montana Oil and Gas Conservation Division.

Mooers, A. O., L. R. Prugh, M. Festa-Bianchet, and J. A. Hutchings. 2007. Biases in legal listing under Canadian endangered species legislation. *Conservation Biology* 21:572–5.

Mooney, H. A., R. N. Mack, J. A. McNeely, L. E. Neville, P. J. Schei, and J. K. Waage. 2005. *Invasive alien species: A new synthesis*. Washington, DC: Island Press.

Morgantini, L. E. 1985. Ungulate encounters with construction materials during the building of an underground gas pipeline in western Alberta. *Alces* 21:215–30.

Morin, N., ed. 1995. *Vascular plants of the United States*. Washington, DC: U.S. Department of the Interior, National Biological Service.

Morse, L. E., J. T. Kartesz, and L. S. Kutner. 1995. Native vascular plants. Pages 205–209 in E. T. LaRoe, G. S. Farris, C. E. Puckett, P. D. Doran, and M. J.

Mac, eds. *Our living resources: A report to the nation on the distribution, abundance, and health of U.S. plants, animals and ecosystems.* Washington, DC: U.S. Department of the Interior, National Biological Service.

Mortberg, U. M., B. Balfors, and W. C. Knol. 2007. Landscape ecological assessment: A tool for integrating biodiversity issues in strategic environmental assessment and planning. *Journal of Environmental Management* 82:457–70.

Murcia, C. 1995. Edge effects in fragmented forests: Implications for conservation. *Trends in Ecology and Evolution* 10:58–62.

Murphy, M. A., J. S. Evans, and A. Storfer. 2010. Quantifying *Bufo boreas* connectivity in Yellowstone National Park with landscape genetics. *Ecology* 91:252–61.

Murray, L. D., and L. B. Best. 2003. Short-term bird response to harvesting switchgrass for biomass in Iowa. *Journal of Wildlife Management* 67:611–21.

Nams, V. O., and M. Bourgeois. 2004. Fractal analysis measures habitat use at different spatial scales: An example with American marten. *Canadian Journal of Zoology* 82:1738–47.

Nassauer, J. I., and R. C. Corry. 2004. Using normative scenarios in landscape ecology. *Landscape Ecology* 19:343–56.

National Academy of Sciences (NAS). 2008. *Environmental impacts of wind-energy projects.* Washington, DC: National Academies Press.

National Academy of Sciences Committee on Onshore Oil and Gas Leasing. 1989. *Land use planning and oil and gas leasing on onshore federal lands.* Washington, DC: National Academies Press.

National Agricultural Statistics Service. 2007. *USDA forecasts record-setting corn crop for 2007.* U.S. Department of Agriculture. www.nass.usda.gov/Newsroom/2007/08_10_2007.asp. Retrieved March 3, 2009.

National Atlas of the United States. 2006. *Federal lands of the United States: National atlas of the United States.* Reston, VA: U.S. Geological Survey.

National Environmental Policy Task Force. 2003. Report to the Council on Environmental Quality 45. nepa.gov/ntf/report/finalreport.pdf. Retrieved April 4, 2008.

National Oceanic and Atmospheric Administration. 1999. *Discounting and the treatment of uncertainty in natural resource damage assessment.* Technical paper 99-1. Silver Spring, MD: National Oceanic and Atmospheric Administration.

National Petroleum Council. 2007. *Facing the hard truths about energy: A comprehensive view to 2030 of global oil and natural gas.* Washington, DC: National Petroleum Council.

National Research Council. 2001. *Compensating for wetland losses under the Clean Water Act.* Washington, DC: National Academies Press.

———. 2003. *Cumulative environmental effects of oil and gas activities on Alaska's north slope.* Washington, DC: National Academies Press.

National Wind Coordinating Committee (NWCC). 2004. *Wind turbine interactions with birds and bats: A summary of research results and remaining questions. Fact sheet*, 2nd ed. www.nationalwind.org/publications/wildlifewind.aspx. Retrieved January 17, 2009.

Nature Conservancy, The. 2000. *Conservation by design: A framework for mission success*. Arlington, VA: The Nature Conservancy.

Naugle, D. E., C. L. Aldridge, B. L. Walker, T. E. Cornish, B. J. Moynahan, M. J. Holloran, K. Brown, et al. 2004. West Nile virus: Pending crisis for greater sage-grouse. *Ecology Letters* 7:704–13.

Naugle, D. E., K. E. Doherty, B. L. Walker, M. J. Holloran, and H. E. Copeland. 2011. Energy development and sage-grouse. Number 21 in S. T. Knick and J. W. Connelly, eds. Greater sage-grouse: Ecology and conservation of a landscape species and its habitats. *Studies in Avian Biology*. sagemap.wr.usgs.gov/monograph.aspx. Retrieved September 10, 2010.

Naylor, L. M., M. J. Wisdom, and R. G. Anthony. 2009. Behavioral responses of North American elk to recreational activity. *Journal of Wildlife Management* 73:328–38.

Nebraska Game and Parks Commission (NGPC). 2009. *Location of sharp-tailed grouse and greater prairie-chicken display grounds in relation to NPPD Ainsworth Wind Energy Facility: 2006–2009*. Lincoln: NGPC.

Nellemann, C., and R. D. Cameron. 1996. Effects of petroleum development on terrain preferences of calving caribou. *Arctic* 49:23–8.

———. 1998. Cumulative impacts of an evolving oil-field complex on the distribution of calving caribou. *Canadian Journal of Zoology* 76:1425–30.

Nellemann, C., I. Vistnes, P. Jordhoy, and O. Strand. 2001. Winter distribution of wild reindeer in relation to power lines, roads and resorts. *Biological Conservation* 101:351–60.

Nellemann, C., I. Vistnes, P. Jordhoy, O. Strand, and A. Newton. 2003. Progressive impact of piecemeal infrastructure development on wild reindeer. *Biological Conservation* 113:307–17.

Neufeld, L. 2006. *Spatial dynamics of wolves and woodland caribou in an industrial forest landscape in west-central Alberta*. M.S. thesis, University of Alberta, Edmonton.

Newsom, D., V. Bahm, and B. Cashore. 2006. Does forest certification matter? An analysis of operation-level changes required during the SmartWood certification process in the United States. *Forest Policy and Economics* 9:197–208.

Newson, A. C. 2001. The future of natural gas exploration in the foothills of the western Canadian Rocky Mountains. *The Leading Edge* 20:74–9.

Nie, M. 2003. *Beyond wolves: The politics of wolf recovery and management*. Minneapolis: University of Minnesota Press.

———. 2008. The underappreciated role of regulatory enforcement in natural resource conservation. *Policy Sciences* 41:139–64.

Nielsen, S. E., S. Herrero, M. S. Boyce, R. D. Mace, B. Benn, M. L. Gibeau, and S. Jevons. 2004. Modelling the spatial distribution of human-caused grizzly bear mortalities in the Central Rockies ecosystem of Canada. *Biological Conservation* 120:101–13.

Nielsen, S. E., G. B. Stenhouse, H. L. Beyer, F. Huettmann, and M. S. Boyce. 2008. Can natural disturbance-based forestry rescue a declining population of grizzly bears? *Biological Conservation* 141:2193–207.

Nielsen, S. E., G. B. Stenhouse, and M. S. Boyce. 2006. A habitat-based framework for grizzly bear conservation in Alberta. *Biological Conservation* 130:217–29.

Nitschke, C. R. 2008. The cumulative effects of resource development on biodiversity and ecological integrity in the Peace-Moberly region of Northeast British Columbia, Canada. *Biodiversity and Conservation* 17:1715–40.

Noble, B. F. 2002. The Canadian experience with SEA and sustainability. *Environmental Impact Assessment Review* 22:3–16.

Nocera, J. J., G. J. Parsons, G. R. Milton, and A. H. Fredeen. 2005. Compatibility of delayed cutting regime with bird breeding and hay nutritional quality. *Agriculture, Ecosystems and Environment* 107:245–53.

Noel, L. E., K. R. Parker, and M. A. Cronin. 2004. Caribou distribution near an oilfield road on Alaska's North Slope, 1978–2001. *Wildlife Society Bulletin* 32:757–71.

———. 2006. Response to Joly et al. (2006): A reevaluation of caribou distribution near an oilfield road on Alaska's North Slope. *Wildlife Society Bulletin* 34:870–3.

Northeast Sage-Grouse Working Group. 2006. *Northeast Wyoming sage-grouse conservation plan*. gf.state.wy.us/wildlife/wildlife_management/sagegrouse/Northeast/NEConsvPlan.pdf.

Northwest Wildlife Consultants, Inc., and Western EcoSystems Technology, Inc. 2007. *Avian and bat monitoring report for the Klondike II Wind Power Project, Sherman County, Oregon*. Prepared for PPM Energy, Portland, OR.

Noss, R. F., C. Carroll, K. Vance-Borland, and G. Wuerthner. 2002. A multicriteria assessment of the irreplaceability and vulnerability of sites in the Greater Yellowstone Ecosystem. *Conservation Biology* 16:895–908.

Noss, R. F., and A. Y. Cooperrider. 1994. *Saving nature's legacy: Protecting and restoring biodiversity*. Washington, DC: Island Press.

Novacek, M. J., and E. E. Cleland. 2001. The current biodiversity extinction event: Scenarios for mitigation and recovery. *Proceedings of the National Academy of Sciences of the United States of America* 98:5466–70.

O'Connell, T. J., L. E. Jackson, and R. P. Brooks. 2000. Bird guilds as indicators of ecological condition in the central Appalachians. *Ecological Applications* 10:1706–21.

Ogle, S. M., W. A. Reiners, and K. G. Gerow. 2003. Impacts of exotic annual brome grasses (*Bromus* spp.) on ecosystem properties of northern mixed grass prairie. *American Midland Naturalist* 149:46–58.

Olson, G. 1981. *Effects of seismic exploration on summering elk in the Two Medicine-*

Badger Creek Area, northcentral Montana. Helena: Montana Fish, Wildlife and Parks.

Opler, P. A. 1978. Insects of American chestnut: Possible importance and conservation concern. Pages 83–85 in J. McDonald, W. L., F. C. Cech, J. Luchock, and C. Smith, eds. *Proceedings of the American Chestnut Symposium.* Morgantown: West Virginia University.

Orloff, S., and A. Flannery. 1992. *Wind turbine effects on avian activity, habitat use, and mortality in Altamont Pass and Solano County Wind Resource Areas, 1989–1991.* Final report P700-92-001 to Alameda, Contra Costa, and Solano counties and the California Energy Commission, Sacramento, California. Tiburon, CA: Biosystems Analysis, Inc.

Osborn, R. G., C. D. Dieter, K. F. Higgins, and R. E. Usgaard. 1998. Bird flight characteristics near wind turbines in Minnesota. *American Midland Naturalist* 139:29–38.

Osko, T. J., M. N. Hiltz, R. J. Hudson, and S. M. Wasel. 2004. Moose habitat preferences in response to changing availability. *Journal of Wildlife Management* 68:576–84.

Otis, D. L., and G. C. White. 1999. Autocorrelation of location estimates and the analysis of radiotracking data. *Journal of Wildlife Management* 63:1039–44.

Oyler-McCance, S. J., and T. W. Quinn. 2011. Molecular insights into the biology of greater sage-grouse. Number 7 in S. T. Knick and J. W. Connelly, eds. Greater sage-grouse: Ecology and conservation of a landscape species and its habitats. *Studies in Avian Biology.* sagemap.wr.usgs.gov/monograph.aspx. Retrieved September 23, 2010.

Panjabi, A. O., E. H. Dunn, P. J. Blancher, W. C. Hunter, B. Altman, J. Bart, C. J. Beardmore, et al. 2005. *The Partners in Flight handbook on species assessment,* Version 2005. Partners in Flight Technical Series No. 3. www.rmbo.org/pubs/downloads/Handbook2005.pdf. Retrieved September 13, 2010.

Parker, K. L. 2003. Advances in the nutritional ecology of cervids at different scales. *Ecoscience* 10:395–411.

Parker, K. L., P. S. Barboza, and M. P. Gillingham. 2009. Nutrition integrates environmental responses of ungulates. *Functional Ecology* 23:57–69.

Parker, K. L., P. S. Barboza, and T. R. Stephenson. 2005. Protein conservation in female caribou (*Rangifer tarandus*): Effects of decreasing diet quality during winter. *Journal of Mammalogy* 86:610–22.

Parmenter, R. R., J. A. MacMahon, M. E. Waaland, M. M. Stuebe, P. Landres, and C. M. Crisafulli. 1985. Reclamation of surface coal mines in western Wyoming for wildlife habitat: A preliminary analysis. *Reclamation and Revegetation Research* 4:93–115.

Parsons, J. J. 1972. Spread of African pasture grasses to American tropics. *Journal of Range Management* 25:12–7.

Patterson, R. L. 1952. *The sage grouse in Wyoming.* Denver, CO: Sage.

Pease, C. M., and D. J. Mattson. 1999. Demography of the Yellowstone grizzly bears. *Ecology* 80:957–75.

Peterson, G. D., G. S. Cumming, and S. R. Carpenter. 2003. Scenario planning: A tool for conservation in an uncertain world. *Conservation Biology* 17:358–66.

Phillips, G. E., and A. W. Alldredge. 2000. Reproductive success of elk following disturbance by humans during calving season. *Journal of Wildlife Management* 64:521–30.

Phillips, S. J., R. P. Anderson, and R. E. Schapire. 2006. Maximum entropy modeling of species geographic distributions. *Ecological Modelling* 190:231–59.

Pigliucci, M. 2002. Are ecology and evolutionary biology "soft" sciences? *Annales Zoologici Fennici* 39:87–98.

Pimentel, D., R. Zuniga, and D. Morrison. 2005. Update on the environmental and economic costs associated with alien-invasive species in the United States. *Ecological Economics* 52:273–88.

Pinay, G., A. Fabre, P. Vervier, and F. Gazelle. 1992. Control of C, N, P distribution in soils of riparian forests. *Landscape Ecology* 6:121–32.

Piorkowski, M. D. 2006. *Breeding bird habitat use and turbine collisions of birds and bats located at a wind farm in Oklahoma mixed-grass prairie*. M.S. thesis, Oklahoma State University, Stillwater.

Pitman, J. C., C. A. Hagen, R. J. Robel, T. M. Loughin, and R. D. Applegate. 2005. Location and success of lesser prairie-chicken nests in relation to vegetation and human disturbance. *Journal of Wildlife Management* 69:1259–69.

Pocewicz, A., M. Nielsen-Pincus, C. S. Goldberg, M. H. Johnson, P. Morgan, J. E. Force, L. P. Waits, and L. Vierling. 2008. Predicting land use change: Comparison of models based on landowner surveys and historical land cover trends. *Landscape Ecology* 23:195–210.

Polasky, S. 2008. Why conservation planning needs socioeconomic data. *Proceedings of the National Academy of Sciences of the United States of America* 105:6505–6.

Pontius, G. R., and J. Malanson. 2005. Comparison of the structure and accuracy of two land change models. *International Journal of Geographical Information Science* 19:243–65.

Possingham, H. P., I. R. Ball, and S. Andelman. 2000. Mathematical methods for identifying representative reserve networks. Pages 291–305 in S. Ferson and M. Burgman, eds. *Quantitative methods for conservation biology*. New York: Springer.

Powell, J. 2003. *Distribution, habitat use patterns, and elk response to human disturbance in the Jack Marrow Hills, Wyoming*. M.S. thesis, University of Wyoming, Laramie.

Powell, R. A. 2000. Animal home ranges and territories and home range estimators. Pages 65–110 in L. Boitani and T. K. Fuller, eds. *Research techniques in animal ecology: Controversies and consequences*. New York: Columbia University Press.

Preisler, H. K., A. A. Ager, and M. J. Wisdom. 2006. Statistical methods for analysing responses of wildlife to human disturbance. *Journal of Applied Ecology* 43:164–72.

Pressey, R. L., and M. C. Bottrill. 2008. Opportunism, threats, and the evolution of systematic conservation planning. *Conservation Biology* 22:1340–5.

Pressey, R. L., H. P. Possingham, and J. R. Day. 1997. Effectiveness of alternative heuristic algorithms for identifying indicative minimum requirements for conservation reserves. *Biological Conservation* 80:207–19.

Pritchard, D. 1993. Towards sustainability in the planning process: The role of EIA. *ECOS: A Review of Conservation* 14:3–15.

Pruett, C. L., M. A. Patten, and D. H. Wolfe. 2009a. Avoidance behavior by prairie grouse: Implications for development of wind energy. *Conservation Biology* 23:1253–59.

——. 2009b. It's not easy being green: Wind energy and a declining grassland bird. *BioScience* 59:257–62.

Pulliam, H. R. 1996. Sources and sinks: Empirical evidence and population consequences. Pages 45–70 in O. E. Rhodes, R. K. Chesser, and M. H. Smith, eds. *Population dynamics in ecological space and time.* Chicago: University of Chicago Press.

Purcell, K. L., and J. Verner. 1998. Density and reproductive success of California towhees. *Conservation Biology* 12:442–50.

Pyare, S., S. Cain, D. Moody, C. Schwartz, and J. Berger. 2004. Carnivore recolonisation: Reality, possibility and a non-equilibrium century for grizzly bears in the southern Yellowstone ecosystem. *Animal Conservation* 7:71–7.

Pyke, D. A. 2011. Restoring and rehabilitating sagebrush habitats. Number 24 in S. T. Knick and J. W. Connelly, eds. Greater sage-grouse: Ecology and conservation of a landscape species and its habitats. *Studies in Avian Biology.* sagemap.wr.usgs.gov/monograph.aspx. Retrieved September 23, 2010.

Pysek, P., and D. M. Richardson. 2007. Traits associated with invasiveness in alien plants: Where do we stand? Pages 97–125 in W. Nentwig, ed. *Biological invasions.* New York: Springer.

Pysek, P., D. M. Richardson, and M. Williamson. 2004. Predicting and explaining plant invasions through analysis of source area floras: Some critical considerations. *Diversity and Distributions* 10:179–87.

Quinonez-Pinon, R., A. Mendoza-Duran, and C. Valeo. 2007. Design of an environmental monitoring program using NDVI and cumulative effects assessment. *International Journal of Remote Sensing* 28:1643–64.

Rabin, L. A., R. G. Coss, and D. H. Owings. 2006. The effects of wind turbines on antipredator behavior in California ground squirrels (*Spermophilus beecheyi*). *Biological Conservation* 131:410–20.

Rahel, F. J. 2002. Homogenization of freshwater faunas. *Annual Review of Ecology and Systematics* 33:291–315.

Raithel, J. D., M. J. Kauffman, and D. H. Pletscher. 2007. Impact of spatial and temporal variation in calf survival on the growth of elk populations. *Journal of Wildlife Management* 71:795–803.

Ramp, D., V. K. Wilson, and D. B. Croft. 2006. Assessing the impacts of roads in peri-urban reserves: Road-based fatalities and road usage by wildlife in the Royal National Park, New South Wales, Australia. *Biological Conservation* 129:348–59.

Randall, J. M. 1996. Weed control for the preservation of biological diversity. *Weed Technology* 10:370–83.

Ratti, J. T., and K. P. Reese. 1988. Preliminary test of the ecological trap hypothesis. *Journal of Wildlife Management* 52:484–91.

Reed, J. C., and C. A. Bush. 2007. *About the geologic map in the* National Atlas of the United States of America. Washington, DC: U.S. Geological Survey. nationalatlas.gov. Retrieved March 3, 2008.

Reid, W. V., and K. R. Miller. 1989. *Keeping options alive: The scientific basis for conservation biodiversity*. Washington, DC: World Resources Institute.

Reijnen, R., R. Foppen, and G. Veenbaas. 1997. Disturbance by traffic of breeding birds: Evaluation of the effect and considerations in planning and managing road corridors. *Biodiversity and Conservation* 6:567–81.

Reimers, E., S. Eftestol, and J. E. Colman. 2003. Behavior responses of wild reindeer to direct provocation by a snowmobile or skier. *Journal of Wildlife Management* 67:747–54.

Rejmanek, M., D. M. Richardson, S. I. Higgins, M. J. Pitcairn, and E. Grotkopp. 2005. Ecology of invasive plants: State of the art. Pages 104–61 in H. A. Mooney, R. N. Mack, J. A. McNeely, L. E. Neville, P. J. Schei, and J. K. Waage, eds. *Invasive alien species: Searching for solutions*. Washington, DC: Island Press.

Remes, V. 2000. How can maladaptive habitat choice generate source–sink population dynamics? *Oikos* 91:579–82.

Renewable Fuels Association. 2009. *U.S. cellulosic ethanol projects under development and construction*. www.ethanolrfa.org/page/-/rfa-association-site/Outlook/CurrentAdvancedCelluloseBiofuelsProjects2-25-10.pdf. Retrieved February 1, 2009.

REN21. 2008. *Renewables 2007 global status report*. Washington, DC: Worldwatch Institute, and Paris: REN21 Secretariat.

Retief, F., C. Jones, and S. Jay. 2008. The emperor's new clothes: Reflections on strategic environmental assessment (SEA) practice in South Africa. *Environmental Impact Assessment Review* 28:504–14.

Reynolds, R. E., T. L. Shaffer, C. R. Loesch, and R. R. Cox, Jr. 2006. The Farm Bill and duck production in the Prairie Pothole Region: Increasing the benefits. *Wildlife Society Bulletin* 34:963–74.

Rhymer, J. M., and D. Simberloff. 1996. Extinction by hybridization and introgression. *Annual Review of Ecology and Systematics* 27:83–109.

Ribic, C. A., R. R. Koford, J. R. Herkert, D. H. Johnson, N. D. Niemuth, D. E. Naugle, K. K. Bakker, D. W. Sample, and R. B. Renfrew. 2009. Area sensitivity in North American grassland birds: Patterns and processes. *Auk* 126:233–44.

Rice, C. A., M. S. Ellis, and J. H. Bullock, Jr. 2000. *Water co-produced with coalbed methane in the Powder River Basin, Wyoming: Preliminary composition data.* Open file report 00-372. Denver, CO: U.S. Geological Survey.

Rich, A. C., D. S. Dobkin, and L. J. Niles. 1994. Defining forest fragmentation by corridor width: The influence of narrow forest-dividing corridors on forest-nesting birds in southern New Jersey. *Conservation Biology* 8:1109–21.

Rich, T. D., C. J. Beardmore, H. Berlanga, P. J. Blancher, M. S. W. Bradstreet, G. S. Butcher, D. W. Demarest, et al. 2005. *Partners in Flight North American landbird conservation plan.* Ithaca, NY: Cornell Lab of Ornithology.

Richardson, D. M., N. Allsopp, C. M. D'Antonio, S. J. Milton, and M. Rejmanek. 2000. Plant invasions: The role of mutualisms. *Biological Reviews* 75:65–93.

Richardson, D. M., and P. Pysek. 2006. Plant invasions: Merging the concepts of species invasiveness and community invasibility. *Progress in Physical Geography* 30:409–31.

Riebsame, W. E. 2001. Geographies of the new West. Pages 43–50 in P. Brick, D. Snow, and S. Van de Wetering, eds. *Across the great divide: Explorations in collaborative conservation and the American West.* Washington, DC: Island Press.

Rittel, H. W. J., and M. M. Webber. 1973. Dilemmas in a general theory of planning. *Policy Sciences* 4:155–69.

Roane, M. K., G. J. Griffin, and J. R. Elkins. 1986. *Chestnut blight, and other* Endothia *diseases, and the genus* Endothia. St. Paul, MN: American Phytopathological Society Press.

Robbins, M. B., and B. C. Dale. 1999. Sprague's pipit (*Anthus spragueii*). In A. Poole and F. Gill, eds. *The Birds of North America* Number 439. Philadelphia: The Birds of North America, Inc.

Robbins, M. B., A. T. Peterson, and M. A. Ortega-Huerta. 2002. Major negative impacts of early intensive cattle stocking on tallgrass prairies: The case of the greater prairie chicken (*Tympanuchus cupido*). *North American Birds* 56:239–44.

Robel, R. J., J. J. A. Harrington, C. A. Hagen, J. C. Pitman, and R. R. Reker. 2004. Effect of energy development and human activity on the use of sand sagebrush habitat by lesser prairie-chickens in southwestern Kansas. *Transactions of the North American Wildlife and Natural Resources Conference* 69:251–66.

Roberts, P. D. 2004. *The end of oil: On the edge of a perilous new world.* New York: Houghton Mifflin.

Robinson, T. W. 1965. *Introduction, spread, and aerial extent of saltcedar (*Tamarix*) in the western states.* Professional paper 491-A. Washington, DC: U.S. Geological Survey.

Roedenbeck, I. A., L. Fahrig, C. S. Findlay, J. E. Houlahan, J. A. G. Jaeger, N. Klar, S. Kramer-Schadt, and E. A. van der Grift. 2007. The Rauischholzhausen agenda for road ecology. *Ecology and Society* 12:11.

Romero, L. M. 2004. Physiological stress in ecology: Lessons from biomedical research. *Trends in Ecology and Evolution* 19:249–55.

Rosentreter, R. 1994. Displacement of rare plants by exotic grasses. U.S. Department of Agriculture, Forest Service General Technical Report INT-0748-1209.

Ross, W. A. 1998. Cumulative effects assessment: Learning from Canadian case studies. *Impact Assessment and Project Appraisal* 16:267–76.

Rost, G. R., and J. A. Bailey. 1979. Distribution of mule deer and elk in relation to roads. *Journal of Wildlife Management* 43:634–41.

Rotenberry, J. T., M. A. Patten, and K. L. Preston. 1999. Brewer's sparrow (*Spizella breweri*). In A. Poole, and F. Gill, eds. *The Birds of North America* Number 390. Philadelphia: The Birds of North America, Inc.

Rothley, K. D. 2001. Manipulative, multi-standard test of a white-tailed deer habitat suitability model. *Journal of Wildlife Management* 65:953–63.

Rowcliffe, J. M., E. de Merode, and G. Cowlishaw. 2004. Do wildlife laws work? Species protection and the application of a prey choice model to poaching decisions. *Proceedings of the Royal Society of London Series B-Biological Sciences* 271:2631–6.

Rowland, J. 2008. *Ecosystem impacts of historical shallow gas wells within the CFB Suffield National Wildlife Area*. Unpublished report, Department of National Defence, Ottawa.

Rowland, M. M., M. J. Wisdom, B. K. Johnson, and J. G. Kie. 2000. Elk distribution and modeling in relation to roads. *Journal of Wildlife Management* 64:672–84.

Ruhl, J. B. 2005. The disconnect between environmental assessment and adaptive management. *Trends* 36:6.

———. 2008. Adaptive management for natural resources: Inevitable, impossible, or both? *Rocky Mountain Law Institute* 54:11.01.

Russell, D. E., R. G. White, and C. J. Daniel. 2005. *Energetics of the porcupine caribou herd: A computer simulation model*. Ottawa: Canadian Wildlife Service.

Saab, V. 1999. Importance of spatial scale to habitat use by breeding birds in riparian forests: A hierarchical analysis. *Ecological Applications* 9:135–51.

Sadar, M. H., D. R. Cressman, and D. C. Damman. 1995. Cumulative effects assessment: The development of practical frameworks. *Impact Assessment* 13:4.

Saher, D. J., and F. K. A. Schmiegelow. 2006. Movement pathways and critical habitat: Selection by mountain caribou during spring migration. *Rangifer* 16:33–45.

Sallabanks, R., E. B. Arnett, and J. M. Marzluff. 2000. An evaluation of research on the effects of timber harvest on bird populations. *Wildlife Society Bulletin* 28:1144–55.

Samarakoon, M., and J. S. Rowan. 2008. A critical review of environmental impact statements in Sri Lanka with particular reference to ecological impact assessment. *Environmental Management* 41:441–60.

Samson, F., and F. Knopf. 1994. Prairie conservation in North America. *BioScience* 44:418–21.

Sansonetti, T. L., and W. R. Murray. 1990. A primer on the Federal Onshore Oil and Gas Leasing Reform Act of 1987 and its regulations. *Land and Water Law Review* 25:375–416.

Sather, N. 1992. *Element stewardship abstract for* Lonicera japonica *(Japanese honeysuckle)*. Arlington, VA: The Nature Conservancy.

Sauer, J. R., J. E. Hines, and J. Fallon. 2008. *The North American breeding bird survey results and analysis 1966–2007*. Version 5.15.2008. www.pwrc.usgs.gov. Retrieved January 10, 2009.

Sawyer, H., M. J. Kauffman, and R. M. Nielson. 2009. Influence of well pad activity on winter habitat selection patterns of mule deer. *Journal of Wildlife Management* 73:1052–61.

Sawyer, H., F. Lindzey, and D. McWhirter. 2002. Potential effects of oil and gas development on mule deer and pronghorn populations in western Wyoming. *Transactions of the North American Wildlife and Natural Resources Conference* 67:350–65.

———. 2005a. Mule deer and pronghorn migration in western Wyoming. *Wildlife Society Bulletin* 33:1266–73.

Sawyer, H., R. Nielson, D. Stickland, and L. McDonald. 2005b. *Mule deer study (Phase II): Long-term monitoring plan to assess potential impacts of energy development on mule deer in the Pinedale Anticline Project Area*. Western Ecosystems Technology, Inc., Cheyenne, WY.

Sawyer, H., R. M. Nielson, F. G. Lindzey, L. Keith, J. H. Powell, and A. A. Abraham. 2007. Habitat selection of Rocky Mountain elk in a nonforested environment. *Journal of Wildlife Management* 71:868–74.

Sawyer, H., R. M. Nielson, F. Lindzey, and L. L. McDonald. 2006. Winter habitat selection of mule deer before and during development of a natural gas field. *Journal of Wildlife Management* 70:396–403.

Schaefer, J. A. 2003. Long-term range recession and the persistence of caribou in the taiga. *Conservation Biology* 17:1435–9.

Schaefer, J. A., and F. Messier. 1995. Habitat selection as a hierarchy: The spatial scales of winter foraging by muskoxen. *Ecography* 18:333–44.

Schaid, T. A., D. W. Uresk, W. L. Tucker, and R. L. Linder. 1983. Effects of surface mining on the vesper sparrow in the northern Great Plains. *Journal of Range Management* 36:500–3.

Schlaepfer, M. A., M. C. Runge, and P. W. Sherman. 2002. Ecological and evolutionary traps. *Trends in Ecology and Evolution* 17:474–80.

Schmidt, W. 1989. Plant dispersal by motor cars. *Vegetatio* 80:147–52.

Schneider, D. C. 2001. The rise of the concept of scale in ecology. *BioScience* 51:545–53.

Schneider, R. R., G. Hauer, W. L. Adamowicz, and S. Boutin. 2010. Triage for conserving populations of threatened species: The case of woodland caribou in Alberta. *Biological Conservation* 143:1603–11.

Schneider, R. R., J. B. Stelfox, S. Boutin, and S. Wasel. 2003. Managing the cumulative impacts of land uses in the Western Canadian Sedimentary Basin: A modeling approach. *Conservation Ecology* 7:8.

Schroeder, M. A., C. L. Aldridge, A. D. Apa, J. R. Bohne, C. E. Braun, S. D. Bunnell, J. W. Connelly, et al. 2004. Distribution of sage-grouse in North America. *Condor* 106:363–76.

Schroeder, M. A., J. R. Young, and C. E. Braun. 1999. Greater sage-grouse (*Centrocercus urophasianus*). In A. Poole, and F. Gill, eds. *The Birds of North America* Number 425. Philadelphia: The Birds of North America, Inc.

Schwartz, C. C., and A. W. Franzmann. 1989. Bears, wolves, moose, and forest succession, and some management considerations on the Kenai Peninsula, Alaska. *Alces* 25:1–11.

Schwartz, C. C., M. A. Haroldson, G. C. White, R. B. Harris, S. Cherry, K. A. Keating, D. Moody, and C. Servheen. 2006. Temporal, spatial, and environmental influences on the demographics of grizzly bears in the Greater Yellowstone Ecosystem. *Wildlife Monographs* 161.

Seip, D. R. 1992. Factors limiting woodland caribou populations and their interrelationships with wolves and moose in southeastern British Columbia. *Canadian Journal of Zoology* 70:1494–1503.

Seip, D. R., C. J. Johnson, and G. S. Watts. 2007. Displacement of mountain caribou from winter habitat by snowmobiles. *Journal of Wildlife Management* 71:1539–44.

Senft, R. L., M. B. Coughenour, D. W. Bailey, L. R. Rittenhouse, O. E. Sala, and D. M. Swift. 1987. Large herbivore foraging and ecological hierarchies. *BioScience* 37:789–99.

Shaffer, J. A., and D. H. Johnson. 2008. Displacement effects of wind developments on grassland birds in the northern Great Plains. Pages 57–61 in *Proceedings of wind wildlife research meeting VII*. Washington, DC: National Wind Coordinating Collaborative.

Sherry, E., and H. Myers. 2002. Traditional environmental knowledge in practice. *Society and Natural Resources* 15:345–58.

Shifley, S. R., F. R. Thompson, W. D. Dijak, and Z. F. Fan. 2008. Forecasting landscape-scale, cumulative effects of forest management on vegetation and wildlife habitat: A case study of issues, limitations, and opportunities. *Forest Ecology and Management* 254:474–83.

Simberloff, D., and P. Stiling. 1996. Risks of species introduced for biological control. *Biological Conservation* 78:185–92.

Sinclair, I. C. 2000. Better laws equal better environment: The role of the environmental lawyer in the reconstruction and modernisation of the water sector in private sector participation. *Water, Air, and Soil Pollution* 123:353–60.

Slade, N. A., and S. M. Blair. 2000. An empirical test of using counts of individuals captured as indices of population size. *Journal of Mammalogy* 81:1035–45.

Smallwood, K. S., and B. Karas. 2009. Avian and bat fatality rates at old-generation and repowered wind turbines in California. *Journal of Wildlife Management* 73:1062–71.

Smallwood, K. S., and C. G. Thelander. 2004. *Bird mortality at the Altamont Pass Wind Resource Area: March 1998–September 2001*. Golden, CO: BioResource Consultants.

Snow, D. 2001. Coming home: An introduction to collaborative conservation. Pages 1–12 in P. Brick, D. Snow, and S. Van de Wetering, eds. *Across the great divide: Explorations in collaborative conservation and the American West*. Washington, DC: Island Press.

Soderman, T. A. 2006. Treatment of biodiversity issues in impact assessment of electricity power transmission lines: A Finnish case review. *Environmental Impact Assessment Review* 26:319–38.

Song, S., ed. 2002. *The ecological basis for stand management: A summary and synthesis of ecological responses to wildfire and harvesting in boreal forests*. Vegreville: Alberta Research Council.

Sorensen, T., P. D. McLoughlin, D. Hervieux, E. Dzus, J. Nolan, B. Wynes, and S. Boutin. 2008. Determining sustainable levels of cumulative effects for boreal caribou. *Journal of Wildlife Management* 72:900–5.

Sovacool, B. K. 2009. Contextualizing avian mortality: A preliminary appraisal of bird and bat fatalities from wind, fossil-fuel, and nuclear electricity. *Energy Policy* 37:2241–8.

Spies, T. A., K. N. Johnson, K. M. Burnett, J. L. Ohmann, B. C. McComb, G. H. Reeves, P. Bettinger, J. D. Kline, and B. Garber-Yonts. 2007. Cumulative ecological and socioeconomic effects of forest policies in coastal Oregon. *Ecological Applications* 17:5–17.

Starfield, A. M. 1997. A pragmatic approach to modeling for wildlife management. *Journal of Wildlife Management* 61:261–70.

Stephens, S. E., J. A. Walker, D. R. Blunck, A. Jayaraman, D. E. Naugle, J. K. Ringelman, and A. J. Smith. 2008. Predicting risk of habitat conversion in native temperate grasslands. *Conservation Biology* 22:1320–30.

Stiver, S. J., A. D. Apa, J. R. Bohne, S. D. Bunnell, P. A. Deibert, S. C. Gardner, M. A. Hilliard, C. W. McCarthy, and M. A. Schroeder. 2006. *Greater sage-grouse comprehensive conservation strategy*. Unpublished report, Western Association of Fish and Wildlife Agencies, Cheyenne, WY.

St-Laurent, M. H., J. Ferron, S. Hache, and R. Gagnon. 2008. Planning timber harvest of residual forest stands without compromising bird and small mammal communities in boreal landscapes. *Forest Ecology and Management* 254:261–75.

St-Laurent, M. H., J. Ferron, C. Hins, and R. Gagnon. 2007. Effects of stand structure and landscape characteristics on habitat use by birds and small mammals in managed boreal forest off eastern Canada. *Canadian Journal of Forest Research* 37:1298–309.

Stohlgren, T. J., D. T. Barnett, and J. Kartesz. 2003. The rich get richer: Patterns of plant invasions in the United States. *Frontiers in Ecology and the Environment* 1:11–4.

Stone, K. 2007. Mineral and metal commodity reviews: Coal. Report catalogue number M38-5/56E-PDF.

Storfer, A., M. A. Murphy, J. S. Evans, C. S. Goldberg, S. Robinson, S. F. Spear, R. Dezzani, E. Delmelle, L. Vierling, and L. P. Waits. 2007. Putting the "landscape" in landscape genetics. *Heredity* 98:128–42.

Stuart-Smith, A. K., C. J. A. Bradshaw, S. Boutin, D. M. Hebert, and A. B. Rippin. 1997. Woodland caribou relative to landscape patterns in northeastern Alberta. *Journal of Wildlife Management* 61:622–33.

Sutter, G. C., and R. M. Brigham. 1998. Avifaunal and habitat changes resulting from conversion of native prairie to crested wheat grass: Patterns at songbird community and species levels. *Canadian Journal of Zoology* 76:869–75.

Sutter, G. C., S. K. Davis, and D. C. Duncan. 2000. Grassland songbird abundance along roads and trails in southern Saskatchewan. *Journal of Field Ornithology* 71:110–6.

Swenson, D. P., and R. F. Ambrose. 2007. A spatial analysis of cumulative habitat loss in Southern California under the Clean Water Act Section 404 program. *Landscape and Urban Planning* 82:41–55.

Swenson, J. E., C. A. Simmons, and C. D. Eustace. 1987. Decrease of sage grouse *Centrocercus urophasianus* after ploughing of sagebrush steppe. *Biological Conservation* 41:125–32.

Tack, J. E. 2009. *Sage-grouse and the human footprint: Implications for conservation of small and declining populations*. M.S. thesis, University of Montana, Missoula.

Taylor, A. R., and R. L. Knight. 2003. Wildlife responses to recreation and associated visitor perceptions. *Ecological Applications* 13:951–63.

ten Kate, K., J. Bishop, and R. Bayon. 2004. Biodiversity offsets: Views, experience, and the business case. International Union for Conservation of Nature, Gland, Switzerland and Cambridge, United Kingdom, and Insight Investment, London.

Tester, J. W., B. J. Anderson, A. S. Batchelow, D. D. Blackwell, R. DiPippo, E. M. Drake, J. Garnish, et al. 2006. *The future of geothermal energy: Impact of enhanced geothermal systems (EGS) on the United States in the 21st century*. Cambridge: Massachusetts Institute of Technology.

Theobald, D. M., and N. T. Hobbs. 1998. Forecasting rural land-use change: A comparison of regression- and spatial transition-based models. *Geographical and Environmental Modelling* 2:65–82.

Thiel, D., S. Jenni-Eiermann, V. Braunisch, R. Palme, and L. Jenni. 2008. Ski tourism affects habitat use and evokes a physiological stress response in capercaillie *Tetrao urogallus*: A new methodological approach. *Journal of Applied Ecology* 45:845–53.

Thorne, J. H., S. Y. Gao, A. D. Hollander, J. A. Kennedy, M. McCoy, R. A. Johnston, and J. F. Quinn. 2006. Modeling potential species richness and urban buildout to identify mitigation sites along a California highway. *Transportation Research Part D-Transport and Environment* 11:277–91.

Thorne, J. H., P. R. Huber, E. H. Girvetz, J. Quinn, and M. C. McCoy. 2009. Integration of regional mitigation assessment and conservation planning. *Ecology and Society* 14:47.

Thrower, J. 2006. Adaptive management and NEPA: How a non-equilibrium view of ecosystems mandates flexible regulation. *Ecology Law Quarterly* 3:871–84.

Thuiller, W., D. M. Richardson, M. Rouget, S. Proches, and J. R. U. Wilson. 2006. Interactions between environment, species traits, and human uses describe patterns of plant invasions. *Ecology* 87:1755–69.

Tierney, R. 2007. *Buffalo Gap I Wind Farm avian mortality study: February 2006–January 2007*. Final report TRC Report Number 110766-C-01 prepared for AES SeaWest, Inc., Albuquerque, NM.

Tilman, D. 1999. The ecological consequences of changes in biodiversity: A search for general principles. *Ecology* 80:1455–74.

Timoney, K., and P. Lee. 2001. Environmental management in resource-rich Alberta, Canada: First world jurisdiction, third world analogue? *Journal of Environmental Management* 63:387–405.

Tisdell, C. A. 1995. Issues in biodiversity conservation including the role of local communities. *Environmental Conservation* 22:216–22.

Toepfer, J. E., and W. L. Vodehnal. 2009. *Greater prairie chickens: Grasslands and vertical structures*. Presentation at the 28th meeting of the Prairie Grouse Technical Council, Portales, NM.

Trail, P. W. 2006. Avian mortality at oil pits in the United States: A review of the problem and efforts for its solution. *Environmental Management* 38:532–44.

TRC Environmental Corporation. 2008. *Post-construction avian and bat fatality monitoring and grassland bird displacement surveys at the Judith Gap Wind Energy Project, Wheatland County, Montana*. Prepared for Judith Gap Energy, LLC, Chicago, IL. TRC Environmental Corporation, Laramie, WY. TRC Project 51883-01 (112416).

Trombulak, S. C., and C. A. Frissell. 2000. Review of ecological effects of roads on terrestrial and aquatic communities. *Conservation Biology* 14:18–30.

Turner, B. L., E. F. Lambin, and A. Reenberg. 2007. The emergence of land change science for global environmental change and sustainability. *Proceedings of the National Academy of Sciences of the United States of America* 104:20666–71.

Turner, M. G., V. H. Dale, and R. H. Gardner. 1989. Predicting across scales: Theory development and testing. *Landscape Ecology* 3:245–52.

Turner, M. G., R. H. Gardner, and R. V. O'Neill. 2001. *Landscape ecology in theory and practice: Pattern and process*. New York: Springer.

Tyler, N. J. C. 1991. Short-term behavioural responses of Svalbard reindeer *Rangifer tarandus platyrhynchus* to direct provocation by a snowmobile. *Biological Conservation* 56:179–94.

Tyser, R. W., and C. A. Worley. 1992. Alien flora in grasslands adjacent to road and trail corridors in Glacier National Park, Montana (USA). *Conservation Biology* 6:253–62.

Underwood, A. J. 1997. *Experiments in ecology: Their logical design and interpretation using analysis of variance*. Cambridge: Cambridge University Press.

Unsworth, J. W., D. F. Pac, G. C. White, and R. M. Bartmann. 1999. Mule deer survival in Colorado, Idaho, and Montana. *Journal of Wildlife Management* 63:315–26.

URS Corporation, W. P. Erickson, and L. Sharp. 2005. Phase 1 and Phase 1A avian mortality monitoring report for 2004–2005 for the Smud Solano Wind Project. Prepared for Sacramento Municipal Utility District, Sacramento, California by Western EcoSystems Technology, Inc., Cheyenne, WY, and Lynn Sharp, environmental consultant.

U.S. Department of Agriculture. 2007. *Environmental management system implementation*. www.fs.fed.us/ems/includes/ems_all_employees_letter.pdf. Retrieved November 11, 2009.

U.S. Department of Energy. 2008a. *20% energy by 2030: Increasing wind energy's contribution to U.S. electricity supply*. Washington, DC: Office of Science and Technical Information.

———. 2008b. *Renewable energy data book*. Washington, DC: Office of Science and Technical Information.

U.S. Department of Energy and U.S. Bureau of Land Management. 2008. *Solar energy development programmatic environmental impact statement*. Washington, DC: U.S. Department of Energy and BLM.

U.S. Department of the Interior. 1997. *White River Resource Area approved resource management plan*. Cheyenne, WY: Bureau of Land Management.

———. 2006. *Record of decision for the Jonah Infill Drilling Project Environmental Impact Statement, Sublette County, Wyoming*. Cheyenne, WY: U.S. Bureau of Land Management.

———. 2007. *Adaptive management implementation policy manual*. Environmental Quality Programs Part 522. Washington, DC: Bureau of Land Management.

———. 2008. National Integrated Lands System (NILS) database. Washington, DC: U.S. Bureau of Land Management.

———. 2009a. News release: Secretary Salazar and Senator Reid announce 'fast-track' initiatives for solar energy development on western lands.

———. 2009b. Solar energy development programmatic EIS. Notice of extension of public comment period for programmatic environmental impact statement to develop and implement agency-specific programs for solar energy development. *Federal Register* 74:37051.

U.S. Department of the Interior, Agriculture, and Energy. 2008. *Inventory of onshore federal oil and natural gas resources and restrictions to their development:*

Phase III inventory—Onshore United States. BLM/WO/GI-03/002+3100/REV08. Washington, DC: U.S. Departments of the Interior, Agriculture, and Energy.

U.S. Environmental Protection Agency (EPA). 2009. *EPA proposes new regulations for the National Renewable Fuel Standard Program for 2010 and beyond*. EPA-420-F-09-023. www.epa.gov/OMS/renewablefuels/420f09023.htm. Retrieved June 3, 2009.

U.S. Fish and Wildlife Service (USFWS). 2003. *Interim guidelines to avoid and minimize wildlife impacts from wind turbines*. Washington, DC: U.S. Fish and Wildlife Service.

———. 2004. *Prairie grouse leks and wind turbines: U.S. Fish and Wildlife Service justification for a 5-mile buffer from leks; additional grassland songbird recommendations*. Unpublished briefing paper.

———. 2008. Review of native species that are candidates for listing as endangered or threatened: Annual notice of findings on resubmitted petitions; annual description of progress on listing actions; proposed rule. *Federal Register* 75:13910–14014.

U.S. Geological Survey. 2008. *Assessment of moderate- and high-temperature geothermal resources of the United States*. Report factsheet 2008-3082.

U.S. Institute for Environmental Conflict Resolution. 2005. *National Environmental Conflict Resolution Advisory Committee final report*. Tucson, AZ: Morris K. Udall Foundation.

Usgaard, R. E., D. E. Naugle, R. G. Osborn, and K. F. Higgins. 1997. Effects of wind turbines on nesting raptors at Buffalo Ridge in southwestern Minnesota. *South Dakota Academy of Science* 76:113–7.

van Dyke, F., and W. C. Klein. 1996. Response of elk to installation of oil wells. *Journal of Mammalogy* 77:1028–41.

Van Horn, M. A., and T. M. Donovan. 1994. Ovenbird (*Seiurus aurocapilla*). In A. Poole and F. Gill, eds. *The Birds of North America* Number 88. Philadelphia: The Birds of North America, Inc.

Ventyx Energy. 2009. EV energy and EV power map (database). Boulder, CO: Ventyx.

Vickery, P. D. 1996. Grasshopper sparrow (*Ammodramus savannarum*). In A. Poole, and F. Gill, eds. *The Birds of North America* Number 239. Philadelphia: The Birds of North America, Inc.

Vistnes, I., and C. Nellemann. 2001. Avoidance of cabins, roads, and power lines by reindeer during calving. *Journal of Wildlife Management* 65:915–25.

———. 2008. The matter of spatial and temporal scales: A review of reindeer and caribou response to human activity. *Polar Biology* 31:399–407.

Vistnes, I., C. Nellemann, P. Jordhoy, and O. Strand. 2001. Wild reindeer: Impacts of progressive infrastructure development on distribution and range use. *Polar Biology* 24:531–7.

———. 2004. Effects of infrastructure on migration and range use of wild reindeer. *Journal of Wildlife Management* 68:101–8.

Vitousek, P. M. 1990. Biological invasions and ecosystem processes: Towards an integration of population biology and ecosystem studies. *Oikos* 57:7–13.

Vitousek, P. M., C. M. Dantonio, L. L. Loope, M. Rejmanek, and R. Westbrooks. 1997a. Introduced species: A significant component of human-caused global change. *New Zealand Journal of Ecology* 21:1–16.

Vitousek, P. M., H. A. Mooney, J. Lubchenco, and J. M. Melillo. 1997b. Human domination of Earth's ecosystems. *Science* 277:494–9.

Vitousek, P. M., L. R. Walker, L. D. Whiteaker, D. Muellerdombois, and P. A. Matson. 1987. Biological invasion by *Myrica faya* alters ecosystem development in Hawaii. *Science* 238:802–4.

Von der Lippe, M., and I. Kowarik. 2007. Long-distance dispersal of plants by vehicles as a driver of plant invasions. *Conservation Biology* 21:986–96.

Vors, L. S., J. A. Schaefer, B. A. Pond, A. R. Rodgers, and B. R. Patterson. 2007. Woodland caribou extirpation and anthropogenic landscape disturbance in Ontario. *Journal of Wildlife Management* 71:1249–56.

Walker, B. G., P. D. Boersma, and J. C. Wingfield. 2005. Physiological and behavioral differences in magellanic penguin chicks in undisturbed and tourist-visited locations of a colony. *Conservation Biology* 19:1571–7.

Walker, B. L., and D. E. Naugle. 2011. West Nile virus ecology in sagebrush habitat and impacts on greater sage-grouse populations. Number 10 in S. T. Knick and J. W. Connelly, eds. Greater sage-grouse: Ecology and conservation of a landscape species and its habitats. *Studies in Avian Biology.* sagemap.wr.usgs.gov/monograph.aspx. Retrieved September 10, 2010.

Walker, B. L., D. E. Naugle, and K. E. Doherty. 2007a. Greater sage-grouse population response to energy development and habitat loss. *Journal of Wildlife Management* 71:2644–54.

———. 2007b. West Nile virus and greater sage-grouse: Estimating infection rate in a wild population. *Avian Diseases* 51:691–6.

Walker, B. L., D. E. Naugle, K. E. Doherty, and T. E. Cornish. 2004. Outbreak of West Nile virus in greater sage-grouse and guidelines for monitoring, handling, and submitting dead birds. *Wildlife Society Bulletin* 32:1000–6.

Walker, J., S. E. Stephens, N. D. Niemuth, C. R. Loesch, R. E. Reynolds, J. S. Gleason, and M. A. Erickson. 2008. *Assessing potential impacts of wind energy development on abundance of breeding duck pairs in the Prairie Pothole Region of North and South Dakota.* Unpublished report, Ducks Unlimited, Bismarck, ND.

Wallace, L. L., M. G. Turner, W. H. Romme, R. V. Oneill, and Y. G. Wu. 1995. Scale of heterogeneity of forage production and winter foraging by elk and bison. *Landscape Ecology* 10:75–83.

Wallgren, M., C. Skarpe, R. Bergstrom, K. Danell, L. Granlund, and A. Bergstrom. 2009. Mammal community structure in relation to disturbance and resource gradients in southern Africa. *African Journal of Ecology* 47:20–31.

Walter, W. D., J. D. M. Leslie, and J. A. Jenks. 2009. Response of Rocky Mountain elk (*Cervus elaphus*) to windpower development. *American Midland Naturalist* 156:363–75.

Walters, C. 1986. *Adaptive management of renewable resources*. New York: Macmillan.

Walters, C. J., and C. S. Holling. 1990. Large-scale management experiments and learning by doing. *Ecology* 71:2060–8.

Ward, L. 1986. Displacement of elk related to seismograph activity in south-central Wyoming. Pages 246–54 in *Issues and technology in the management of impacted western wildlife: Proceedings of a national symposium*. Boulder, CO: Thorne Ecological Institute.

Wärnbäck, A., and T. Hilding-Rydevik. 2009. Cumulative effects in Swedish EIA practice: Difficulties and obstacles. *Environmental Impact Assessment Review* 29:107–15.

Wasser, S. K., K. Bevis, G. King, and E. Hanson. 1997. Noninvasive physiological measures of disturbance in the northern spotted owl. *Conservation Biology* 11:1019–22.

Webb, N. F., M. Hebblewhite, and E. H. Merrill. 2008. Statistical methods for identifying wolf kill sites using global positioning system locations. *Journal of Wildlife Management* 72:798–807.

Weber, E. P. 2000. A new vanguard for the environment: Grass-roots ecosystem management as a new environmental movement. *Society and Natural Resources* 13:237–59.

Weclaw, P., and R. J. Hudson. 2004. Simulation of conservation and management of woodland caribou. *Ecological Modelling* 177:75–94.

Weisenberger, M. E., P. R. Krausman, M. C. Wallace, D. W. DeYoung, and O. E. Maughan. 1996. Effects of simulated jet aircraft noise on heart rate and behavior of desert ungulates. *Journal of Wildlife Management* 60:52–61.

West, D. W., N. Ling, B. J. Hicks, L. A. Tremblay, N. D. Kim, and M. R. v. d. Heuvel. 2006. Cumulative impacts assessment along a large river, using brown bullhead catfish (*Ameiurus nebulosus*) populations. *Environmental Toxicology and Chemistry* 25:1868–80.

West, N. E. 1983. Western intermountain sagebrush steppe. Pages 351–69 in N. E. West, ed. *Temperate deserts and semi-deserts*. Amsterdam: Elsevier Scientific.

Westbrooks, R. G. 2004. New approaches for early detection and rapid response to invasive plants in the United States. *Weed Technology* 18:1468–71.

Western Governors' Association. 2009. *Western renewable energy zones—Phase 1 report: Mapping concentrated, high quality resources to meet demand in the Western Interconnection's distant markets*. Washington, DC: Western Governors' Association.

Wheatley, M., and C. Johnson. 2009. Factors limiting our understanding of ecological scale. *Ecological Complexity* 6:150–9.

Wheelwright, N. T., and J. D. Rising. 2008. Savannah sparrow (*Passerculus sandwichensis*). In A. Poole, and F. Gill, eds. *The Birds of North America* Number 45. Philadelphia: The Birds of North America, Inc.

Whisenant, S. G. 1990. *Changing fire frequencies on Idaho's Snake River Plains: Ecological and management implications*. Forest Service General Technical Report INT-276. Washington, DC: U.S. Department of Agriculture.

White, P., and J. T. Kerr. 2006. Contrasting spatial and temporal global change impacts on butterfly species richness during the 20th century. *Ecography* 29:908–18.

Whitmore, R. C., and G. A. Hall. 1978. The response of passerine birds to a new resource: Reclaimed surface mines in West Virginia. *American Birds* 32:6–9.

Wiens, J. A. 1989. Spatial scaling in ecology. *Functional Ecology* 3:385–97.

Wilcove, D. S., D. Rothstein, J. Dubow, A. Phillips, and E. Losos. 1998. Quantifying threats to imperiled species in the United States. *BioScience* 48:607–15.

Wilderness Society, The. 2005. *Oil and gas on public lands: An overview*. Washington, DC: The Wilderness Society.

Wilkinson, J. B., J. M. McElfish, Jr., R. Kihslinger, R. Bendick, and B. A. McKenney. 2009. *The next generation of mitigation: Linking current and future mitigation programs with state wildlife action plans and other state and regional plans*. Washington, DC: Environmental Law Institute and The Nature Conservancy.

Williams, B. K., R. C. Szaro, and C. D. Shapiro. 2009. *Adaptive management: The U.S. Department of the Interior Technical Guide*. Washington, DC: Adaptive Management Working Group, U.S. Department of the Interior.

Wilson, E. O. 1992. *The diversity of life*. Cambridge, MA: Harvard University Press.

Wilson, K., R. L. Pressey, A. Newton, M. Burgman, H. Possingham, and C. Weston. 2005. Measuring and incorporating vulnerability into conservation planning. *Environmental Management* 35:527–43.

With, K. A., A. W. King, and W. E. Jensen. 2008. Remaining large grasslands may not be sufficient to prevent grassland bird declines. *Biological Conservation* 141:3152–67.

Wittmer, H. U., A. R. E. Sinclair, and B. N. McLellan. 2005. The role of predation in the decline and extirpation of woodland caribou. *Oecologia* 144:257–67.

Wondolleck, J. M., and S. L. Yaffee. 2000. *Making collaboration work: Lessons from innovation in natural resource management*. Washington, DC: Island Press.

World Bank. 2007. *Global economic prospects 2007: Managing the next wave of globalization*. Washington, DC: World Bank.

Wray, T., K. A. Strait, and R. C. Whitmore. 1982. Reproductive success of grassland sparrows on a reclaimed surface mine in West Virginia. *Auk* 99:157–64.

Wyoming Fish and Game Department. 2009. *Recommendations for development of oil and gas resources within important wildlife habitats*. Unpublished report. Cheyenne, WY.

Yahner, R. H. 2008. Bird responses to a managed forested landscape. *Wilson Journal of Ornithology* 120:897–900.

Yamasaki, S. H., R. Duchesneau, F. Doyon, J. S. Russell, and T. Gooding. 2008. Making the case for cumulative impacts assessment: Modelling the potential impacts of climate change, harvesting, oil and gas, and fire. *Forestry Chronicle* 84:349–68.

Yoccoz, N. G. 1991. Use, overuse, and misuse of significance tests in evolutionary biology and ecology. *Bulletin of the Ecological Society of America* 72:106–11.

Yoder J. M., D. A. Swanson, and E. A. Marschall. 2004. The cost of dispersal: Predation as a function of movement in ruffed grouse. *Behavioral Ecology* 15:469–76.

You, M. Q. 2008. Moratorium on EIA approvals: China's new environmental law enforcement tool. *Natural Resources Journal* 48:163–87.

Young, D. P. J., W. P. Erickson, and J. P. Eddy. 2005a. *Mountain plover (*Charadrius montanus*) surveys, Foote Creek Rim Wind Plant, Carbon County, Wyoming, 1995–2005*. Unpublished report to PacifiCorp and SeaWest Windpower, Inc. Western EcoSystems Technology, Inc., Cheyenne, WY.

Young, D. P. J., W. P. Erickson, R. E. Good, M. D. Strickland, and G. D. Johnson. 2003. *Avian and bat mortality associated with the initial phase of the Foote Creek Rim Windpower Project, Carbon County, Wyoming*. Technical report for Eurus Energy America Corporation and the Combine Hills Technical Advisory Committee, Umatilla County, OR. Prepared by Western EcoSystems Technology, Inc., Cheyenne, WY, and Northwest Wildlife Consultants, Inc., Pendleton, OR.

Young, D. P., Jr., W. P. Erickson, J. Jeffrey, K. Bay, and M. Bourassa. 2005b. *Eurus Combine Hills Turbine Ranch. Phase I post-construction wildlife monitoring final report*. Western EcoSystems Technology, Incorporated, Cheyenne, WY, and Northwest Wildlife Consultants, Inc., Pendleton, OR.

Young, D. P. J., W. P. Erickson, J. Jeffrey, and V. K. Poulton. 2007. *Puget Sound Energy Hopkins Ridge Wind Project Phase 1 post-construction avian and bat monitoring first annual report*. Report for Puget Sound Energy, Dayton, WA and Hopkins Ridge Wind Project Technical Advisory Committee, Columbia County, WA by Western EcoSystems Technology, Inc., Cheyenne, WY.

Young, D. P. J., J. Jeffrey, W. P. Erickson, K. Bay, and V. K. Poulton. 2006. *Eurus Combine Hills Turbine Ranch. First annual report, phase 1 post construction wildlife monitoring to Eurus Energy America Corporation, San Diego, California, and the Combine Hills Technical Advisory Committee, Umatilla County, Oregon*. Western EcoSystems Technology, Inc., Cheyenne, WY, and Northwest Wildlife Consultants, Inc., Pendleton, OR.

Zalatan, R., A. Gunn, and G. H. R. Henry. 2006. Long-term abundance patterns of barren-ground caribou using trampling scars on roots of *Picea mariana* in the Northwest Territories, Canada. *Arctic Antarctic and Alpine Research* 38:624–30.

Zeiler, H. P., and V. Grünschachner-Berger. 2009. Impact of wind power plants on black grouse, *Lyrurus tetrix* in Alpine regions. *Folia Zoologica* 58:173–82.

Zellmer, S. B. 2000. The virtues of "command and control" regulation: Barring exotic species from aquatic ecosystems. *University of Illinois Law Review* 2000:1233–86.

Zhu, D., and J. Ru. 2008. Strategic environmental assessment in China: Motivations, politics, and effectiveness. *Journal of Environmental Management* 88:615–26.

Zou, L., S. N. Miller, and E. T. Schmidtmann. 2006. Mosquito larval habitat mapping using remote sensing and GIS: Implications of coalbed methane development and West Nile virus. *Journal of Medical Entomology* 43:1034–41.

Erin M. Bayne is an associate professor in the Department of Biological Sciences at the University of Alberta. His research is focused on understanding how human land use decisions influence animal behavior and how changes in behavior influence population and community processes. He has worked in the boreal forest for 16 years on a wide variety of species and has more than forty scientific publications. His current emphasis is on estimating the relative impacts of forestry and energy sector development on boreal forest songbirds.

Melinda Harm Benson is an assistant professor in the Department of Geography at the University of New Mexico. Her research and teaching focus on environment and natural resource management challenges, with a particular emphasis on emerging trends in environmental governance, including adaptive management. She received her J.D. from the University of Idaho. Before academic life, she worked first as a lobbyist and then as an attorney representing conservation groups on environment and natural resource issues in the Intermountain West.

Erin Bergquist is a plant ecologist at the environmental consulting firm AECOM Environment. She holds an M.S. in ecology from Colorado State University in Fort Collins. Erin's graduate research focused on studying invasive weed species in and around coal bed natural gas developments. Her current work includes applying weed management and planning to large-scale industrial projects and permitting.

Mark S. Boyce is a professor in the Department of Biological Sciences at the University of Alberta, where he holds the Alberta Conservation Association Chair in Fisheries and Wildlife. He studies population ecology and conservation of a variety of vertebrate species, primarily in the Rocky Mountain region of North America.

Shawn M. Cleveland recently completed his M.S. at the University of Montana and now works for The Nature Conservancy as a wildlife biologist on the Matador Ranch and Grass Bank in eastern Montana. Shawn conducts research on the ranch and uses findings to further his ties with local landowners to achieve landscape conservation.

Holly E. Copeland is a spatial ecologist with The Nature Conservancy in Lander, Wyoming, where her research focuses on sustainable energy development through the use of geographic information systems and modeling tools for mitigation planning and forecasting future impacts of energy development on wildlife. She holds an M.A. in geography from the University of Wyoming.

Alycia W. Crall recently received her Ph.D. at the University of Wisconsin–Madison from the Nelson Institute of Environmental Studies. She has been conducting research on invasive plant species for the past 9 years in the western and midwestern United States. Her primary research interests include invasive species ecology, vegetation monitoring, and citizen science.

Brenda C. Dale is a wildlife biologist for Environment Canada/Canadian Wildlife Service in the Prairie and Northern Region. Brenda has 30 years of experience in monitoring and evaluating the effects of agricultural and industrial practices on grassland birds. Brenda's most recent research provided the basis for her testimony as expert witness for the government of Canada at the 2008 Joint Review Panel Hearings related to EnCana's application to conduct infill drilling within the Canadian Forces Base Suffield National Wildlife Area.

Kevin E. Doherty is a wildlife biologist with the U.S. Fish and Wildlife Service in Bismarck, North Dakota. He holds an M.S. in wildlife conservation from the University of Minnesota and a Ph.D. in wildlife biology from the University of Montana. Kevin has been a leading researcher in sage-grouse landscape ecology and conservation planning. His work has resulted in innovative ways to balance energy development with sage-grouse conservation.

Alison L. Duvall is a conservation practitioner in western Montana and assistant coordinator of the Intermountain West Joint Venture. For 14 years she has worked with nonprofits such as the Blackfoot Challenge, Five Valleys Land Trust, and other community-based organizations and agencies to build private and public partnerships to deliver strategic habitat conservation. Ali holds an M.S. in environmental science from the University of Montana. Her interest is in bridging conservation partnerships, transferring innovations, and building capacity across multiple scales to benefit natural resources and rural communities.

Paul H. Evangelista is a research scientist at the Natural Resource Ecology Laboratory at Colorado State University. He holds an M.S. in forest ecology and a Ph.D. in forest ecosystem management, both from Colorado State University. His research focuses largely on invasive species, but Paul is also recognized for his work on biological soil crusts, fire ecology, East African wildlife, pine bark beetles, and species distribution models.

Mark Hebblewhite is an assistant professor in the Wildlife Biology Program at the University of Montana. Mark's research approach is largely empirical, based on field studies, and makes use of advances in spatial and statistical modeling, including resource selection functions, animal survival analyses, and landscape simulations using geographic information systems. His newest work focuses on the biological response of big game to energy development, with implications for woodland caribou recovery in the Canadian Rockies and northern British Columbia.

Matthew J. Holloran has researched sage-grouse in Wyoming for 15 years. During this time, Matt has served as principal investigator, field supervisor, and collaborator on at least thirteen different research projects addressing various aspects of sage-grouse ecology and management. Matt works as a senior ecologist with Wyoming Wildlife Consultants LLC, based in Laramie, Wyoming.

Chris J. Johnson is an associate professor at the University of Northern British Columbia in Prince George. His research program integrates the disciplines of landscape, wildlife, and conservation ecology. For the past 10 years he has studied the cumulative impacts of human development, focusing on the distributional and population responses of woodland and barren-ground caribou, grizzly bears, wolves, and fisher.

Gregory D. Johnson is an ecologist and senior manager with Western EcoSystems Technology in Cheyenne, Wyoming. He is a certified wildlife biologist and holds an M.S. in zoology and physiology from the University of Wyoming. He has 15 years of experience conducting avian, bat, and other wildlife research associated with wind power developments in sixteen U.S. states and Alberta, Canada, and has authored twenty-seven publications and twenty-seven presentations at scientific meetings on wind power interactions with birds and bats.

Joseph M. Kiesecker is a lead scientist for The Nature Conservancy. He has published more than 100 articles on topics ranging from climate change to the effectiveness of conservation strategies. He holds a Ph.D. in zoology from Oregon State University and has held faculty appointments at Yale

University, Pennsylvania State University, and the University of Wyoming. His recent work includes developing new tools to blend landscape-level planning with mitigation in attempts to better inform land use decisions and is currently testing this process through a series of infrastructure (i.e., energy development and mining) pilot projects.

Bruce A. McKenney is a senior economic advisor for The Nature Conservancy's Conservation Lands program in Washington, D.C. He applies economic and policy analysis to advance biodiversity conservation strategies and currently co-leads the conservancy's Development by Design approach for mitigation planning. He holds an M.S. degree in public policy from Harvard University.

Gregory A. Neudecker is a fish and wildlife biologist for the U.S. Fish and Wildlife Service in Montana. He has worked for the agency since 1987 and has been responsible for establishing conservation focus areas in Montana. His professional interest is linking rural communities with landscape-scale conservation.

Amy Pocewicz is a landscape ecologist with the Wyoming chapter of The Nature Conservancy. She holds an M.S. in remote sensing and a Ph.D. in landscape ecology from the University of Idaho. Her research focuses on the ecological and social implications of landscape change, conservation planning methods, and the effectiveness of conservation strategies. She has coauthored several publications on energy mitigation planning approaches and the potential biological impacts of energy development.

Scott E. Stephens is the director of conservation planning and programs for Ducks Unlimited in Bismarck, North Dakota. His pioneering research highlights the importance of maintaining large and intact glaciated prairie landscapes for waterfowl production in the North American midcontinent.

Martin-Hugues St-Laurent is assistant professor at the Université du Québec à Rimouski, where he teaches and conducts research in animal ecology, wildlife management, and conservation planning. He holds a Ph.D. in biology from the Université du Québec à Montréal. Martin is a landscape ecologist studying the impacts of habitat alteration on mammal conservation. His latest investigations focus on the conservation of woodland caribou, gray wolf, moose, and black bear.

James W. Stutzman is the Montana state coordinator for the U.S. Fish and Wildlife Service's Partners for Fish and Wildlife Program. He holds a B.S. in wildlife biology and an M.S. in environmental education, both from the

University of Minnesota. For more than 20 years he has collaborated with rural communities to implement locally led, landscape-level conservation in priority landscapes.

Jason D. Tack is a research associate at the University of Montana in Missoula. He holds an M.S. in wildlife biology from the University of Montana. Jason has worked extensively on the landscape needs of sage-grouse in the face of human development, and his research focuses on species habitat modeling for conservation planning.

Brett L. Walker is a researcher with the Colorado Division of Wildlife in Grand Junction. He holds an M.S. in avian ecology and a Ph.D. in wildlife biology from the University of Montana and has studied behavior, ecology, demography, and conservation of songbirds, shorebirds, and game birds since 1992. His recent work has contributed to a scientific foundation for greater sage-grouse conservation in response to West Nile virus and oil and gas development.

ABOUT THE EDITOR

David E. Naugle is an associate professor in the Wildlife Biology Program at the University of Montana in Missoula. He completed his Ph.D. at South Dakota State University and spent 3 years at the University of Wisconsin–Stevens Point before moving to Missoula, where he now teaches courses in wildlife habitat and landscape conservation. Dave's applied research focuses on the application of habitat and population models to regional conservation planning for grassland, wetland, and shrub-steppe birds. For 15 years he has studied the ecology and conservation of birds in the western United States and Canada. Highlights of past work include formulating a landscape approach to wetland and grassland bird conservation, assessing effects of wind energy on birds, evaluating conservation benefits of the Conservation Reserve Program, constructing risk models to characterize factors influencing tillage of native rangeland resources, and probing the sociology behind changing land use practices. More recently, Dave has been working with colleagues to assess the vulnerability of prairie wetlands to climate change and to quantify what those changes will mean for mid-continental waterfowl populations. Since 2003, Dave has been evaluating impacts of oil and gas development on sage-grouse populations throughout the West. As part of this work, Dave and his students were the first to discover and quantify mortality of sage-grouse from West Nile virus, an exotic disease of high management concern that is now endemic throughout the West. In 2007, and again in 2008, the Bureau of Land Management in Wyoming and Montana hired Dave as an advisor to help the agency use science to inform their management of wildlife and energy resources on public lands. In 2010, Dave took a sabbatical from the university to help the U.S. Department of Agriculture (Natural Resources Conservation Service) implement their new Sage-Grouse Initiative to benefit wildlife on working ranches in the West. To deliver the initiative, Dave provides planning tools that inform the conservation service about where to apply Farm Bill funding to provide the greatest benefits to sage-grouse populations. Dave was recently named Wildlife Biologist of the Year by the Montana chapter of the Wildlife Society for his leadership in sage-grouse research and conservation.

291

INDEX

Figures/photos/illustrations are indicated by an "f" and tables by a "t".

abiotic conditions, 118
accounting. *See* offset accounting
adaptive management, x, 85, 94, 177, 227; implementation of, 197; law v. practice of, 203; monitoring plan, 197, 207; National Research Council's definition of, 196–197; NEPA and, 204–209; planning and, 208–209; on public lands, energy development and, 196–197; on public lands as new paradigm, 202–210
Adaptive Management Working Group, 205
aerial surveys, 73
aggregated offsets, 172–174
agriculture, 37–38
Alberta (Canada), x, 71, 99, 109
Alberta Energy, 10
ALCES. *See* A Landscape Cumulative Effects Simulator
Altamont Pass Wind Resource Area (APWRA), 133–134, 139, 141
American chestnut tree, 120
American coot, 137
American kestrels, 134, 137
American robin, 101
analysis, 33
animal behavior, impact on individuals and, 41–43
anthropogenic development: community responses to cumulative impacts of, 46–48; population response to cumulative impacts of, 44–46
anthropogenic disturbances, 46–48

APWRA. *See* Altamont Pass Wind Resource Area
Athabasca, 10
avian fatality rates. *See* birds
avoidance behaviors, 34, 42; of birds, 140–141; of elk, 80, 88t, 129; of grassland songbirds, 106, 107t, 108–109; of mule deer, 83, 88t; of northern harriers, 141; of ovenbirds, 99, 100; of pronghorn, 81; roads and, 87–88, 88t; of sage-grouse, 61–63; of songbirds, 110–112; of ungulates, 86, 93, 129; of woodland caribou, 76, 88t

BACI. *See* before-after control-impact design
Baird's sparrow, 106–107, 108, 109
bald eagles, 45
Banff National Park, 47–48
barren-ground caribou, 42, 45, 46, 51, 91
bat fatality rates, 135t–136t, 137–140
bears, 45. *See also specific types*
bedrock, 184–185
before-after control-impact design (BACI), 73; mule deer, 83; on open-country birds, 109; ovenbird experiment, 99–100
big brown bat, 138
biodiesel, 148
biodiversity: invasive species v. native, 120; mitigation hierarchy and, 160f; targets of, 171–172
biofuels, 132, 147–155, 190. *See also* ethanol
biological planning, 217

118; vehicles and, 119; vulnerability of ecosystem v., 118–119
invasive species, 115–116, 159, 226; conservation v., 126–128; cumulative effects of, 126–128; habitats v., 118–119; native biodiversity v., 120; native species v., 118, 120–121, 125t, 126f; traits of, 117–118; water resources v., 119, 122
inverse distance weighted interpolation, 185

Japanese brome, 126
Japanese honeysuckle, 118
Jonah Natural Gas Field, 172, 173f
juniper titmice, 112

Kendall Island Migratory Bird Sanctuary, 108
keystone species, 215
killdeer, 142
kit foxes, 92

Land and Water Conservation Fund, U.S., 180
land leases, 8; coal, 12f–13f, 14; geothermal, 12f, 16f, 18, 22; natural gas, 10, 11f–12f, 38, 186; oil and gas, 10, 11f–13f, 38, 186, 198; oil sands, 12f; oil shale, 12f; solar, 12f, 16f, 17, 21–22; terrestrial ecosystems v., 19t–20t; uranium, 13f, 14; wind, 12f, 15, 16f, 18, 21–22
land management, 8; agencies of, 202–203
land protection, 224–225
land tenure laws, 8, 189–190
land use: planning, 40; wildlife v. regulatory frameworks of, 27–28; wind energy v., 15–16
landowner-led conservation, 219–220, 226
landscape conservation, 64
landscape conservation planning: mitigation hierarchy and, 163–169; mitigation planning at an ecoregional scale, 166; mitigation planning for focal species, 166–169; mitigation

planning for great sage-grouse, 167–168, 191
A Landscape Cumulative Effects Simulator (ALCES), 51
landscape planning, 211–212
Lapland Longspur, 108
laws: 1872 mining, 197; environmental impact, 38–39; federal air and water quality, 190; land tenure, 8. See also legislation
Le Conte's sparrow, 101
leaf litter, 118
leases, 191, 195; BLM influence on, 198–199; contract/provisions of, 198–199; MLA, 197–200; stipulations, 199–200. See also land leases
legislation: alternative, 39; certification systems and, 39–40; enforcement of, 39; environmental impact, 38–39; forest, 39; industrial development v., 40; objectivity of, 41; self-regulation and, 40; variation of, 39. See also laws
licenses, 8
Lincoln's sparrow, 101, 103–104
linear disturbances, 42
liquid petroleum, 190
little brown bat, 138
livestock industry, ix, 153
logging, 47

magnitude of effect, 36–38
mallard, 137
management policy, 93
Mannix, David, 223
Manyberries Oil Field, 59
Marxan algorithm, 171–172, 173f
mechanical control, 128
Melaleuca, 117
migration: of elk, 90; identifying appropriate scale and, 88–91; of mule deer, 90–91; partial, 90; of pronghorn, 90–91; songbird collision during, 96–97; ungulates, 88–91
migratory tree bats, 139–140
mineral development, 186
Mineral Leasing Act of 1920 (MLA), 197–200